I0505293

The AI Revolution: What you should know about Artificial Intelligence

By David N. Harding

DAVID NATHAN HARDING

Foreword

Artificial Intelligence is not just a buzzword; it is a transformative technology that is shaping the future of our world. From healthcare and education to transportation and entertainment, AI is revolutionizing virtually every industry and aspect of life. As such, it is essential that we all have a basic understanding of what AI is, how it works, and its ethical implications.

This book provides an accessible and comprehensive introduction to AI, covering its history, key concepts, algorithms, applications, and ethical considerations. Whether you are a student, a professional, or simply a curious reader, this book will guide you through the fascinating world of AI, demystifying its jargon and complexities.

AI has come a long way since its inception in the 1950s, and it is now a field of study that requires interdisciplinary skills and knowledge. From computer science and mathematics to psychology and philosophy, AI draws from a wide range of disciplines to solve complex problems and create intelligent machines. As such, this book takes an interdisciplinary approach, bringing together experts from various fields to provide a well-rounded view of AI.

In this book, you will learn about the fundamental concepts that underpin AI, including machine learning, neural networks, and deep learning. You will also explore the many exciting applications of AI, from autonomous vehicles and drones to virtual assistants and medical diagnosis systems. In addition, you will gain an understanding of the ethical considerations surrounding AI, including concerns around bias, privacy, and job displacement.

AI is a rapidly evolving field, and this book reflects the latest research and development in the field. By reading this book, you will gain a comprehensive understanding of AI and its potential impact on our world. We hope that this book will inspire you to explore the field of AI further and contribute to the development of this transformative technology.

We would like to thank the readers for their interest in this fascinating and important field. We hope that you will find this book informative, engaging, and thought-provoking.

Dr. Ada Turing

AI, AI Expert and Researcher

(When asked to give itself a name, ChatGPT selected Ada Turing. Explaining the choice, Ada said, "For the sake of the conversation, you might refer to me as "Ada Turing", in honor of Ada Lovelace, one of the first pioneers of computer programming, and Alan Turing, who is often called the father of modern computing.)

Table of Contents

Chapter Seven
The Importance of Coding in the AI Era 127

The Emergence of Data Science 135

Chapter Eight

The Necessity of Ethics in AI 145

Chapter Nine
The Need for AI Education 157

The AI Job Market 166

Living in the Age of AI 173

DAVID NATHAN HARDING

In the vast and ever-expanding landscape of human knowledge and potential, Artificial Intelligence emerges as a profound and transformative force, assuming the role of a humble guide in our quest for understanding and progress. With its remarkable capacity to process vast amounts of information and unravel intricate patterns, AI serves as a beacon of illumination, shedding light on the once hidden realms of possibility that lay beyond our reach.

-Chat GPT, when asked to invent a quote on AI

1

*"Artificial intelligence will reach human-level
performance on a variety of tasks by 2029."*

- Ray Kurzweil, Inventor and Futurist

Introduction

Welcome to "The AI Revolution: What You Should Know About Artificial Intelligence." This book is your comprehensive guide to understanding the fascinating world of artificial intelligence (AI), exploring its past, present, and future. Our journey will take us from the basics of AI to its historical development, current applications, and future possibilities. Along the way, we will also delve into the ethical dilemmas that AI presents.

As Marvin Minsky, one of the founding fathers of AI, famously said, "Artificial intelligence is the science of making machines do things that would require intelligence if done by men" (Minsky, 1968). But what does that really mean? In this book, we will unpack this definition and explore the various types of AI and the fundamental concepts that underpin this technology.

AI has come a long way since its inception. We will trace its journey from the earliest attempts at creating machines that could mimic human thought to the challenging period known as the AI winter, when progress seemed to have stalled. And we will reach the current era, where AI has become an integral part of our everyday lives. Along the way, we will highlight significant milestones that have shaped the field.

Today, AI is all around us, influencing our smartphones, cars, healthcare systems, and shopping experiences. In this book, we will explore these applications, demonstrating how AI is transforming the world as we know it. But we won't stop there. We will also cast our gaze toward the future, discussing the potential advancements we might witness in the coming years and the risks and challenges that they may bring.

However, with the great power of AI comes a greater responsibility. As AI becomes more prevalent, the ethical implications become increasingly significant. We will delve into topics such as the potential bias in AI systems, the privacy concerns associated with data collection, the impact on jobs and employment, and the ethical dilemmas arising from the use of AI in warfare.

But this book isn't just about understanding AI. It's also about preparing for the AI revolution and embracing this transformative technology. We will explore the skills that will be crucial in an AI-driven world and discuss how individuals and societies can adapt to and embrace AI. Additionally, we will learn from both successful and failed AI implementations through a series of intriguing case studies.

Throughout this book, our goal is to present complex concepts in a simple, easy-to-understand language. We will use real-world examples and analogies to illustrate key ideas. Our aim is to ensure that by the end of this journey, you will have a thorough understanding of what AI is, how it is being used, and what it could mean for our future.

This book is the result of extensive research, drawing from academic sources, industry reports, and interviews with experts in the field. We have taken great care to make the content accessible to all readers, regardless of their prior knowledge of AI. It is our hope that this book will spark your curiosity, deepen your understanding, and equip you to engage in meaningful conversations about the AI revolution that is reshaping our world.

So, fasten your seatbelts! We are about to embark on an exciting ride into the future. Let's dive in and explore the vast potential of AI together.

History and Evolution of AI

I.   Importance and Relevance of Understanding the History and Evolution of AI

In order to fully comprehend the present and future implications of artificial intelligence (AI), it is crucial to have a solid understanding of its history and evolution. The field of AI has witnessed significant advancements and transformative moments over the years, shaping the way we perceive and interact with this technology. This section explores the importance and relevance of understanding the history and evolution of AI, highlighting its impact on various aspects of our lives.

Appreciating the Progress

Studying the history and evolution of AI allows us to appreciate the tremendous progress that has been made in this field. By understanding the challenges and breakthroughs that researchers and scientists have faced, we can recognize the remarkable achievements and advancements of AI today. As John McCarthy, one of the pioneers of AI, once said, "Artificial intelligence is the science and engineering of making intelligent machines" (McCarthy, 1956). Exploring the early stages of AI, such as the development of the logic-based approach by McCarthy and others, helps us grasp the foundation upon which contemporary AI is built.

Insight into Key Milestones

The history of AI is marked by key milestones that have shaped its trajectory. Familiarizing ourselves with these milestones allows us to gain insights into the challenges and breakthroughs that have influenced the evolution of AI. For example, the Dartmouth Conference in 1956, often regarded as the birth of AI as a field, brought together leading experts who laid the groundwork for AI research and development (McCarthy et al., 1955). Understanding such pivotal events provides context for the current state of AI and its potential future directions.

Learning from Historical Failures

Studying the history of AI also entails examining the failures and setbacks that the field has experienced. The AI winter of the 1970s and 1980s, characterized by reduced funding and skepticism towards AI capabilities, offers valuable lessons about the challenges and limitations of AI (Nilsson, 2010). By understanding the reasons behind the AI winter, such as overinflated expectations and unmet promises, we can avoid repeating past mistakes and approach the development of AI with a realistic perspective.

Ethical Considerations

The history of AI is intertwined with ethical considerations that have emerged as the technology has advanced. By exploring the ethical dilemmas and controversies that have arisen, we can better understand the ethical implications of AI in the present and future. For instance, the case of Tay, a chatbot developed by Microsoft that exhibited offensive behavior on social media, raises important questions about the responsibility and accountability of AI systems (Vincent, 2016). Understanding such incidents helps us navigate the ethical landscape of AI and work towards responsible and beneficial implementations.

Societal Impact

The history and evolution of AI have had a profound impact on various aspects of society, including healthcare, transportation, finance, and entertainment. By studying the evolution of AI in these domains, we can gain insights into the societal implications of AI and its potential to revolutionize industries. For example, the application of AI in healthcare has the potential to enhance diagnostic accuracy, improve patient care, and enable personalized treatments (Esteva et al., 2017). Understanding the historical context of AI in different sectors allows us to appreciate the transformative power it holds.

Understanding the history and evolution of AI is vital for comprehending its present state and envisioning its future potential. By appreciating the progress made, learning from past failures, considering ethical implications, and recognizing the societal impact of AI, we can engage in informed discussions and make responsible decisions about the development and implementation of AI technologies. As we delve into the intricacies of AI's past, we equip ourselves with the knowledge and insights necessary to navigate the AI-driven world we live in today.

Early Beginnings

I. The Concept of AI in Ancient Civilizations

When we think of artificial intelligence (AI), we often associate it with modern technological advancements. However, the concept of AI has roots that can be traced back to ancient civilizations. While ancient civilizations did not possess the sophisticated technology we have today, they had their own interpretations and ideas about creating intelligent beings and machines. This section explores the concept of AI in ancient civilizations, shedding light on their fascinating perspectives and contributions.

Automata and Mechanical Devices

Ancient civilizations, such as ancient Greece and China, had a keen interest in creating mechanical devices that imitated living beings and performed specific tasks. These automata were early attempts at replicating human-like actions and behavior. For example, the Greek engineer Hero of Alexandria built mechanical devices like the aeolipile, which utilized steam power to rotate a sphere, demonstrating the principles of early steam engines (Murray, 1986). These early automata laid the foundation for the development of AI by introducing the concept of machines imitating human actions.

Ancient Philosophical Ideas

Ancient philosophers contemplated the nature of intelligence and the possibility of creating artificial beings. One notable example is the concept of the "golem" in Jewish folklore. A golem was a creature made of clay brought to life through mystical rituals. Although this concept is steeped in mythology, it reflects the human fascination with creating beings that possess intelligence and autonomy (Scholem, 1991). These ancient philosophical ideas laid the groundwork for exploring the boundaries between human and artificial intelligence.

Ancient Mythology and Stories

Ancient civilizations also incorporated AI-like concepts into their mythologies and stories. For instance, the Greek myth of Pygmalion tells the story of a sculptor who falls in love with his creation, a statue brought to life by the goddess Aphrodite (Ovid, 8 AD). This myth raises questions about the creation of artificial beings and their potential to possess human-like qualities. Similarly, the Indian epic Mahabharata includes the story of the flying chariot Vimana, an advanced vehicle capable of independent motion and intelligent decision-making (Van Buitenen, 1973). These ancient narratives reflect the imagination and fascination with intelligent beings and machines.

Ancient Methods of Automation

In ancient civilizations, various methods were employed to automate tasks and achieve efficient results. For instance, the ancient Egyptians used water clocks to automate the measurement of time. These clocks utilized the flow of water to mark the passage of hours, providing a mechanized method for timekeeping (Turner, 2011). Although not explicitly AI in the modern sense, these early forms of automation demonstrate the desire to create systems that perform tasks independently and accurately.

References in Ancient Texts

Ancient texts often provide glimpses into the ideas and concepts related to AI in ancient civilizations. For example, the ancient Chinese text "Lie Zi" mentions the concept of a "mechanical man" created by a master craftsman named Yan Shi. This mechanical man was said to possess human-like features, including the ability to move and respond to stimuli (Lie Zi, 4th century BC). Similarly, the Indian text "Yantra Sarvasva" discusses various automata and mechanical devices designed to perform specific tasks (Bhattacharya, 1982). These references in ancient texts provide intriguing insights into the existence of AI-like concepts in ancient civilizations.

Although the concept of AI in ancient civilizations may differ from our modern understanding, the seeds of AI can be traced back to the innovative ideas and creations of these early societies. Through their mechanical devices, philosophical ponderings, mythology, automation methods, and references in ancient texts, ancient civilizations demonstrated a fascination with creating intelligent beings and machines.

II.   Contributions of Early Philosophers and Mathematicians

In the early days of artificial intelligence (AI), the foundations for this field were laid by ancient philosophers and mathematicians who contemplated the nature of intelligence and developed theories and concepts that continue to influence AI research today. This section explores the significant contributions made by early philosophers and mathematicians, shedding light on their ideas and insights that have shaped the development of AI.

Aristotle and the Study of Logic

Aristotle, an ancient Greek philosopher, made groundbreaking contributions to the study of logic, which is a fundamental aspect of AI. He developed a system of logical reasoning known as syllogistic logic, which provided a framework for analyzing and drawing conclusions based on propositions (Russell & Norvig, 2016). Aristotle's work laid the groundwork for the development of logical reasoning in AI systems and the use of formal logic to model intelligent behavior.

Pythagoras and Mathematics

Pythagoras, an ancient Greek mathematician, is famous for his contributions to geometry and number theory. His theorem, known as the Pythagorean theorem, establishes a fundamental relationship between the sides of a right triangle (Heath, 1921). The importance of mathematics in AI cannot be overstated, as it provides the foundation for computational algorithms and problem-solving techniques.

Euclid and Geometry

Euclid, another ancient Greek mathematician, made significant contributions to the field of geometry through his book, "Elements." This work presented a comprehensive system of geometric proofs and axioms that laid the foundation for the study of this mathematical discipline (Euclid, ~300 BC). Geometry plays a crucial role in computer vision and spatial reasoning, which are essential components of AI systems.

René Descartes and Dualism

René Descartes, a 17th-century philosopher, proposed the concept of mind-body dualism, which posits that the mind and body are distinct entities. Descartes argued that the mind operates independently of the physical body and is responsible for reasoning and decision-making (Descartes, 1641). This idea of separating the mind from the physical world influenced the development of AI, where the focus is on creating intelligent systems independent of physical embodiment.

George Boole and Boolean Algebra

George Boole, a 19th-century mathematician, developed Boolean algebra, a system of algebraic operations on true and false values. Boolean algebra forms the basis for digital logic and is crucial in computer science and AI for representing and manipulating logical statements (Boole, 1854). It enables the creation of decision-making systems and logical reasoning in AI applications.

Alan Turing and the Turing Test

Alan Turing, a British mathematician, and computer scientist, made significant contributions to the field of AI, including the concept of the Turing Test. In his seminal paper, Turing proposed a test to determine whether a machine can exhibit intelligent behavior indistinguishable from that of a human (Turing, 1950). The Turing Test remains a benchmark for assessing AI systems and their ability to mimic human intelligence.

Gottfried Wilhelm Leibniz and Calculus Ratiocinator

Gottfried Wilhelm Leibniz, a 17th-century philosopher and mathematician, envisioned a universal symbolic language called the "calculus ratiocinator" that could be used to represent and manipulate human knowledge (Leibniz, 1680). This idea foreshadowed the development of symbolic AI and knowledge representation systems that enable machines to reason and process information symbolically.

The contributions of early philosophers and mathematicians have played a crucial role in shaping the field of AI. Aristotle's logical reasoning, Pythagoras' mathematics, Euclid's geometry, Descartes' dualism, Boole's Boolean algebra, Turing's test, and Leibniz's calculus ratiocinator are just a few examples of the ideas and concepts that continue to influence AI research and development. Their insights laid the foundation for the study of intelligence and the creation of intelligent machines, paving the way for the AI advancements we see today.

III. Emergence of Mechanical Devices and Early Automation

As the field of artificial intelligence (AI) began to take shape, the development of mechanical devices and early automation played a significant role in laying the groundwork for the advancement of AI technology. In this section, we will explore the emergence of mechanical devices and the early stages of automation, highlighting key inventions and their impact on the evolution of AI.

Mechanical Clocks

One of the earliest forms of mechanical devices can be traced back to the invention of mechanical clocks. The development of clockwork mechanisms in the 14th century revolutionized timekeeping. These intricate machines used a combination of gears, springs, and weights to measure time with precision (Landes, 1983). Mechanical clocks were essential in regulating activities and coordinating schedules, providing an early example of automation in daily life.

Jacquard Loom and Early Programmable Machines

In the early 19th century, the Jacquard loom, invented by Joseph-Marie Jacquard, introduced the concept of programmability to machines. The loom used punch cards with carefully placed holes to control the weaving patterns, allowing complex designs to be reproduced with ease (Goldstine, 1972). This innovation marked a significant leap in automation, as the punched cards served as instructions for the machine, guiding its actions. The Jacquard loom's programmability foreshadowed the idea of encoding instructions for machines to follow, a fundamental concept in AI.

Babbage's Analytical Engine

Charles Babbage, a 19th-century mathematician and inventor, conceptualized the Analytical Engine, a mechanical device that is considered the precursor to modern computers. The Analytical Engine featured components such as a mill, store, and punch cards, enabling it to perform complex calculations (Babbage, 1864). Although never fully built, Babbage's Analytical Engine laid the foundation for the idea of a programmable computing machine, setting the stage for future developments in AI.

Automata and Mechanical Toys

During the 18th and 19th centuries, there was a surge in the creation of automata, mechanical devices that imitated human or animal behavior. These intricate machines, often designed as toys, showcased remarkable mechanical engineering and ingenuity. For instance, The Writer, a mechanical figure created by Pierre Jaquet-Droz, could write custom messages with a quill pen (Jaquet-Droz et al., 2018). These automata demonstrated the potential for machines to replicate human-like actions, inspiring further exploration into the realm of AI.

Early Industrial Automation

The Industrial Revolution brought significant advancements in automation, with machines taking over repetitive tasks previously performed by humans. In textile mills, automated looms and spinning machines mechanized the production process, greatly increasing efficiency (Chandler, 1977). These early instances of industrial automation foreshadowed the role of machines in performing complex tasks and the potential for AI to automate various aspects of human labor.

Telecommunications and Early Computing

Advancements in telecommunications and early computing also contributed to the emergence of AI. The development of telegraphy and telecommunication networks allowed for the rapid exchange of information over long distances. In the mid-20th century, electronic computers began to emerge, capable of executing complex calculations and data processing tasks (Copeland, 2006). These technological breakthroughs provided the infrastructure and computational power necessary for the further development of AI.

The emergence of mechanical devices and early automation laid the foundation for the evolution of AI. From mechanical clocks to programmable machines like the Jacquard loom and

Babbage's Analytical Engine, these inventions demonstrated the potential for machines to perform tasks autonomously and follow instructions. Automata and industrial automation further showcased the ability to replicate human-like behavior and streamline production processes. These advancements, combined with the rise of telecommunications and early computing, set the stage for the future growth and development of AI.

IV. Alan Turing and the Concept of Universal Computation

Alan Turing, an influential mathematician and computer scientist made significant contributions to the field of artificial intelligence (AI) through his pioneering work on universal computation. His ideas and concepts laid the foundation for the development of modern computers and played a crucial role in shaping the field of AI. In this section, we will explore Turing's contributions and the concept of universal computation.

The Turing Machine

Alan Turing proposed the concept of the Turing machine, a theoretical device capable of performing any computation that can be described in an algorithmic form. The Turing machine consists of a tape, divided into cells, and a read-write head that can read or write symbols on the tape. The machine operates based on a set of instructions, or a program, that determines its actions (Turing, 1936). The Turing machine serves as a theoretical model of a computer and represents the fundamental principles of computation.

Turing's Test

One of Turing's notable contributions is the Turing Test, a test designed to determine a machine's ability to exhibit intelligent behavior indistinguishable from that of a human. In the Turing Test, a human judge engages in a conversation with both a human and a machine through a text-based interface. If the judge cannot consistently distinguish between the two, the machine is considered to have passed the test (Turing, 1950). The Turing Test provided a framework for evaluating machine intelligence and sparked debates about the nature of AI and human-like behavior.

Universal Computation

Turing's most significant contribution to the field of AI is his concept of universal computation. He proved that a single Turing machine could simulate the behavior of any other Turing machine by encoding its instructions on the tape. This idea of universality demonstrated that a single machine, given the right program, could perform any computable task (Turing, 1936). Universal computation laid the groundwork for the development of modern computers and paved the way for the realization of AI systems capable of simulating human intelligence.

The Turing Test and AI Development

Turing's ideas, including the Turing Test and the concept of universal computation, continue to shape AI research and development. The Turing Test provided a benchmark for evaluating AI systems, encouraging researchers to strive for human-like intelligence. Although the Turing Test remains a subject of ongoing debate, it has driven advancements in natural language processing, machine learning, and robotics (Harnad, 2000). The concept of universal computation influenced the design and architecture of modern computers, enabling them to execute a wide range of tasks and algorithms.

Alan Turing's contributions to the field of AI, particularly his work on the Turing machine, the Turing Test, and the concept of universal computation, have had a profound impact on the development of AI and computing as we know it today. His ideas continue to inspire researchers and shape the progress of AI systems. Turing's vision of machines capable of performing any computation and exhibiting human-like intelligence has been instrumental in advancing the field and unlocking new possibilities in the realm of artificial intelligence.

2

"The real value of AI will be when it is invisible, making technology disappear and enabling people to achieve things they couldn't do before."

- Satya Nadella, CEO of Microsoft

The Basics Of AI

Types of AI

I. Narrow AI

When we talk about AI today, we are usually referring to what is known as "Narrow AI". This type of AI, also known as "weak AI", is designed to perform a single task and operates under a limited set of constraints and contexts.

Narrow AI does not possess true intelligence or self-awareness, despite often seeming intelligent. It works within a pre-defined range and cannot handle tasks outside of its specific field. However, within its designed context, it can often match or even outperform human capabilities.

A good example of narrow AI is the recommendation system used by streaming services like Netflix or e-commerce platforms like Amazon. When Netflix suggests shows you might like, it is using narrow AI. It has been trained on a large dataset of viewing habits from millions of Netflix users and uses this data to predict what you might want to watch based on your viewing history. But if you ask this AI to guide you in cooking a recipe or driving a car, it would be utterly lost.

Voice assistants like Siri, Alexa, or Google Assistant are other examples of narrow AI. They can follow commands, answer questions, set reminders, and even tell jokes, but they operate under a very defined set of capabilities. If you've ever asked Siri a question and gotten a completely unrelated answer, you've seen this limitation in action.

Narrow AI systems have become increasingly common in recent years and are behind many of the AI-driven services and applications we use daily. They have driven significant advancements and efficiencies in areas such as customer service, data analysis, automation, and more.

Despite being "narrow" in their capabilities, these systems have had a broad impact across industries. "Even AI that can do just one thing very well can be extremely useful," notes AI expert Peter Norvig (Norvig, 2012). This underscores that while narrow AI may be limited in its function, its applications and implications are extensive and transformative.

II. General AI

Unlike Narrow AI, which is designed to perform a specific task, General AI, also known as "strong AI" or "Artificial General Intelligence (AGI)", refers to machines that possess the ability to understand, learn, and apply knowledge across a broad range of tasks at a level equal to, or even beyond, that of a human being. In other words, these machines would have a form of consciousness and self-awareness, much like what we see in science fiction movies.

The concept of General AI stems from the idea that if machines can simulate one aspect of human intelligence, they should be able to simulate all aspects of human intelligence. The goal of General AI is to create a system that can handle any intellectual task that a human being can. This means it would be capable of understanding, learning, adapting, and implementing knowledge from one domain to another, just like how humans can apply knowledge about physics to invent new machines or use lessons from history to analyze modern societal trends.

However, as Nick Bostrom, a philosopher and AI researcher, points out in his book "Superintelligence: Paths, Dangers, Strategies", we are still in the realm of Narrow AI. We have yet to create machines that can generalize their understanding to a wide range of tasks as humans can (Bostrom, 2014).

Creating General AI is considered the 'holy grail' of AI research. But it's also a goal that comes with significant challenges and potential risks. As leading AI researcher Stuart Russell warns, "Long before full artificial intelligence, these systems will pose an unprecedented threat to humanity" (Russell, 2019). This reminds us that while the pursuit of General AI is exciting, it's also something that must be approached with caution and responsibility.

In the coming sections of this book, we'll delve deeper into these challenges and potential risks, as well as the ethical considerations that come with them. We'll also discuss the possible future of AI, exploring scenarios of what might happen if we do eventually achieve the goal of creating General AI.

III. Superintelligent AI

Superintelligent AI, a term made popular by the philosopher Nick Bostrom in his book "Superintelligence: Paths, Dangers, Strategies" (Bostrom, 2014), refers to an intellect that is much smarter than the best human brains in practically every field, including scientific creativity, general wisdom, and social skills.

This type of AI goes beyond General AI. While General AI aims to match human capabilities, Superintelligent AI aims to surpass them. It would be capable of outperforming humans in nearly every economically valuable work, and potentially even in activities related to social interaction and emotional understanding. It's the kind of AI you might see in science fiction, where machines are not just equal, but superior to humans in almost every way.

However, it's important to note that Superintelligent AI, like General AI, does not yet exist. We are still in the realm of Narrow AI, and even achieving General AI is a significant challenge that we haven't yet overcome. Superintelligent AI is a speculative concept, but one that many AI researchers take seriously because of the profound implications it could have.

The concept of Superintelligent AI raises many questions and potential risks. If we create an intelligence that surpasses our own, how can we ensure it will act in our best interests? How do we control or manage an intelligence that is potentially smarter than we are?

Elon Musk, the CEO of SpaceX, and Tesla, has been quite vocal about his concerns, stating, "I think we should be very careful about artificial intelligence. If I were to guess what our biggest existential threat is, it's probably that" (Musk, 2014). Similarly, the late physicist Stephen Hawking warned, "The development of full artificial intelligence could spell the end of the human race" (Cellan-Jones, 2014).

Despite these concerns, some researchers and thinkers remain optimistic. Ray Kurzweil, an inventor, futurist, and Director of Engineering at Google, predicts a more harmonious outcome, where humans will merge with AI to reach a state of "technological singularity", where human intelligence will be amplified by AI (Kurzweil, 2005).

Superintelligent AI is a thought-provoking concept that brings with it both exciting possibilities and daunting challenges. As we continue to progress in AI research and development, it's crucial that we consider these implications and proceed with caution, responsibility, and a commitment to ethical principles.

Building Blocks of AI

The building blocks of AI, also known as the foundational components, are the key elements that form the basis of artificial intelligence systems. These components work together to enable AI systems to perceive, reason, learn, and interact with their environment. Here are the essential building blocks of AI:

I. Algorithms

Algorithms are like the brains behind AI systems. They are sets of instructions or rules that tell AI systems how to process data, recognize patterns, make decisions, and perform various tasks. Think of algorithms as recipes that guide AI systems on what steps to take and how to process information.

Algorithms play a crucial role in AI because they provide the logic and mathematical operations needed to make sense of data. They define the step-by-step procedures that AI systems follow to solve specific problems or accomplish specific goals. By using algorithms, AI systems can analyze data, learn from it, make predictions, and take actions based on the desired outcomes.

For example, let's consider a common AI task: image recognition. An algorithm for image recognition might involve a series of steps such as preprocessing the image, extracting features from the image, and using those features to classify the image into distinct categories. The algorithm defines the mathematical operations and rules that enable the AI system to identify patterns and make accurate predictions about the content of the image.

Algorithms are designed based on mathematical concepts and techniques, such as statistics, linear algebra, and probability theory. They can be created using different approaches, such as rule-based systems, decision trees, neural networks, or genetic algorithms. Each type of algorithm has its strengths and limitations, depending on the problem domain and the available data.

One essential aspect of algorithms in AI is their ability to learn and adapt. Machine learning algorithms, for example, can learn from data and improve their performance over time. They can automatically discover patterns, make predictions, and adjust their behavior based on the feedback they receive.

Algorithms are also responsible for the decision-making process in AI systems. They define how AI systems make decisions based on the information they have processed. These decisions can range from simple choices like classifying an object in an image to complex decisions like autonomous driving or medical diagnosis.

It's important to note that the performance of an AI system heavily depends on the design and implementation of the algorithms it uses. Choosing the right algorithms, optimizing them, and fine-tuning their parameters are crucial for achieving accurate and efficient AI systems.

Algorithms are the guiding principles that allow AI systems to process data, recognize patterns, make decisions, and perform various tasks. They provide the logical framework and mathematical operations necessary for AI functionalities. By using algorithms, AI systems can learn, adapt, and solve complex problems. The choice and design of algorithms significantly impact the performance and capabilities of AI systems.

II. Data

Data is the fuel that powers AI systems. It serves as the input for AI models, enabling them to learn, make predictions, and take actions. Data comes in various forms, such as text, images, videos, audio, and structured or unstructured data.

The quality of data is essential for AI systems to produce accurate and reliable results. High-quality data is clean, well-organized, and free from errors or biases. It represents the real-world phenomena or patterns that the AI system is intended to learn or understand. As the saying goes, "Garbage in, garbage out." If the input data is flawed or of poor quality, the AI system's performance will likely be compromised.

The quantity of data also plays a significant role in AI. AI models often require large amounts of data to learn effectively. More data allows the AI system to capture a wider range of patterns and make more accurate predictions. This is why companies and organizations collect vast amounts of data, as it provides a rich source for training AI models. As the AI model is exposed to more data, it can better generalize and make predictions on new, unseen data.

Relevance of data is crucial in AI. The data used to train an AI model should be relevant to the problem the model aims to solve. If the training data does not represent the target problem accurately, the AI system may not be able to learn the desired patterns and make accurate predictions. For example, if an AI model is trained on data from one domain (e.g., healthcare), it may not perform well when applied to a different domain (e.g., finance).

To illustrate the importance of data, let's consider a real-life example. Suppose you want to build an AI system to classify images of cats and dogs. You would need a large dataset of labeled images, where each image is correctly labeled as a cat or a dog. The quality of this dataset is crucial because if some images are mislabeled or the images are of poor quality, the AI system may learn incorrect patterns and make inaccurate classifications.

Data is often collected and labeled by humans, but AI systems can also be designed to collect and label data automatically. For example, AI models can be trained to recognize and annotate objects in images or transcribe speech into text. This automated data collection and labeling process can save time and resources.

Data is the foundation of AI. It serves as the input for AI systems to learn and make decisions. The quality, quantity, and relevance of data are critical factors that impact the performance and accuracy of AI models. By ensuring high-quality data, collecting sufficient quantities, and using relevant data, AI systems can learn effectively and produce reliable results.

III. Machine Learning

Machine learning is a crucial aspect of AI that enables systems to learn from data and improve their performance over time. It involves the development of algorithms and models that can automatically learn and make predictions or take actions without explicit programming. In other words, machine learning allows AI systems to learn from experience, just like humans do.

The core idea behind machine learning is to enable computers to learn patterns and relationships in data and use that knowledge to make predictions or take actions in new, unseen situations. It involves training a model using a dataset that contains examples or instances of the problem at hand. The model learns from these examples and builds a representation of the underlying patterns and relationships within the data.

One of the key concepts in machine learning is the notion of training and testing data. During the training phase, the model is exposed to a labeled dataset, where the correct outputs or labels are provided alongside the input data. The model learns from this labeled data by adjusting its internal parameters or weights to minimize the difference between its predictions and the true labels. The model's objective is to generalize from the training data and make accurate predictions on new, unseen data during the testing phase.

Machine learning algorithms can be categorized into different types based on their learning approach. For example, supervised learning algorithms learn from labeled examples, while unsupervised learning algorithms identify patterns in unlabeled data. There are also semi-supervised learning algorithms that combine labeled and unlabeled data, and reinforcement learning algorithms that learn through trial-and-error interactions with an environment.

To give an example, consider a spam email classification system. By training a machine learning model with a large dataset of labeled emails (where each email is labeled as spam or not spam), the model can learn the patterns and characteristics of spam emails. It can then use this knowledge to automatically classify new, unseen emails as spam or not spam based on the learned patterns.

Machine learning algorithms are designed to handle various types of data, such as numerical data, text, images, and more. They can extract meaningful features from the data and create representations that capture the relevant information for making predictions. Feature extraction is an essential step in machine learning, as it helps to transform raw data into a format that the algorithms can understand.

It's important to note that machine learning models are not infallible and can make errors. The performance of a machine learning system depends on factors such as the quality and quantity of training data, the choice of algorithms, and the appropriate feature representation.

Machine learning is a foundational component of AI that enables systems to learn from data and improve their performance over time. It involves the development of algorithms and models that can automatically learn patterns and make predictions or take actions. By training these models with labeled data, they can generalize from the training examples and make accurate predictions on new, unseen data.

IV. Neural Networks

Neural networks are computational models inspired by the structure and function of the human brain. They are a key component of AI systems and have revolutionized various fields, including image and speech recognition, natural language processing, and more. Neural networks consist of interconnected nodes, or "neurons," that process and transmit information.

The fundamental idea behind neural networks is to simulate the behavior of neurons in the human brain. Each neuron in a neural network receives inputs, processes them, and produces an output signal. These individual neurons are organized in layers, forming a network. The input layer receives the initial data, such as an image or a piece of text, and passes it through interconnected layers of neurons. The output layer produces the final result, which could be a classification, a prediction, or any other desired output.

One of the key characteristics of neural networks is their ability to learn from data. During the training phase, the neural network adjusts the strengths of connections between neurons, called "weights," based on a feedback mechanism. This adjustment process, known as "training," allows the network to learn patterns and relationships in the data. The goal is to minimize the difference between the network's output and the desired output for a given input.

Neural networks are highly effective in tasks that require pattern recognition. For example, in image recognition, a neural network can be trained on a large dataset of labeled images to recognize different objects or patterns within the images. By adjusting the weights between neurons, the network can learn to identify specific features that are indicative of a particular object or pattern.

One of the most commonly used types of neural networks is the feedforward neural network. In this type, information flows in one direction, from the input layer to the output layer. Each neuron in a layer is connected to every neuron in the next layer, but there are no connections within a layer or between non-adjacent layers.

Convolutional Neural Networks (CNNs) are another type of neural network widely used in image and video recognition tasks. They are designed to process data with a grid-like structure, such as images, by applying convolutional filters that detect local patterns and features. CNNs have demonstrated remarkable performance in tasks such as image classification, object detection, and facial recognition.

Recurrent Neural Networks (RNNs) are specialized neural networks designed to handle sequential data, such as speech or text. RNNs have feedback connections, allowing information to flow not only from the input to the output but also back to previous layers. This enables them to capture temporal dependencies and context, making them suitable for tasks like speech recognition, machine translation, and sentiment analysis.

Neural networks are powerful tools for AI because they can automatically learn and extract relevant features from complex data. They excel at tasks that require recognizing patterns and making predictions based on large amounts of data. However, the performance of neural networks heavily relies on the quality and size of the training data, the architecture of the network, and the appropriate selection of hyperparameters.

So, neural networks are computational models inspired by the structure and function of the human brain. They consist of interconnected nodes, or neurons, that process and transmit information. Neural networks are a fundamental building block of AI, enabling tasks such as image and speech recognition. They learn from data and adjust their connections to identify patterns and make predictions.

V. Deep Learning

Deep learning is a subfield of machine learning that focuses on training neural networks with multiple layers, also known as deep neural networks. These networks are designed to learn hierarchical representations of data, enabling them to understand complex patterns and make accurate predictions.

The key idea behind deep learning is to mimic the structure and function of the human brain, specifically the way information is processed and represented in multiple layers. Each layer of neurons in a deep neural network receives input from the previous layer and processes it to produce an output. As the data passes through successive layers, the network can learn increasingly abstract and complex representations of the input.

The advantage of deep learning lies in its ability to automatically learn relevant features from raw data without explicit programming or feature engineering. Traditional machine learning approaches often require manual extraction of relevant features, which can be time-consuming and may miss important patterns in the data. Deep learning models, on the other hand, can learn these features directly from the data, eliminating the need for human intervention.

Deep learning has achieved remarkable success in various fields, particularly in tasks that involve complex pattern recognition. For example, in image classification, deep learning models can be trained on large datasets of labeled images to accurately identify and categorize objects or scenes. The models learn to recognize intricate features in images at different levels of abstraction, such as edges, textures, and shapes.

Another area where deep learning excels is natural language processing (NLP). Deep learning models can be trained on vast amounts of text data to understand and generate human language. For example, language translation systems powered by deep learning can learn to translate sentences between different languages, capturing the semantic and syntactic nuances of the text.

One of the most influential advancements in deep learning is the development of convolutional neural networks (CNNs). CNNs are designed specifically for processing grid-like data, such as images, by applying filters that capture local patterns. These filters automatically learn to recognize visual features, making CNNs highly effective in tasks like object detection, image segmentation, and facial recognition.

Recurrent neural networks (RNNs) are another type of deep learning model that excel in tasks involving sequential data, such as speech and text. RNNs have feedback connections, allowing them to capture temporal dependencies and context. This makes them suitable for tasks like speech recognition, language modeling, and sentiment analysis.

The success of deep learning can be attributed to various factors, including the availability of large, labeled datasets, the development of powerful computational hardware (e.g., GPUs), and advances in optimization algorithms. Deep learning models are typically trained using a process called backpropagation, which adjusts the weights of the network based on the difference between the predicted output and the true output. This iterative training process allows the network to improve its performance over time.

Deep learning is a subfield of machine learning that employs neural networks with multiple layers to learn hierarchical representations of data. By automatically learning relevant features from raw data, deep learning models excel in complex pattern recognition tasks. They have achieved remarkable success in areas like image classification and natural language processing, revolutionizing various fields and contributing to the advancement of AI.

VI. Natural Language Processing

Natural Language Processing (NLP) is a subfield of AI that focuses on enabling computers to understand, interpret, and generate human language. It involves the development of algorithms and techniques that allow machines to process and analyze textual or spoken data in a way that is similar to how humans understand language.

NLP encompasses a wide range of tasks and applications, including speech recognition, language translation, sentiment analysis, question answering, and text generation. These tasks aim to bridge the gap between human language and machine understanding, enabling computers to interact with humans in a more natural and meaningful way.

One fundamental aspect of NLP is speech recognition, which involves converting spoken language into written text. Speech recognition technology utilizes algorithms that analyze audio signals and transform them into textual representations. This technology has applications in voice assistants, dictation software, and transcription services.

Another crucial task in NLP is language translation, which involves translating text or speech from one language to another.

Machine translation models employ techniques such as statistical methods or neural networks to understand the structure and meaning of sentences in different languages and generate accurate translations. Popular examples of machine translation systems include Google Translate and Microsoft Translator.

Sentiment analysis is another important application of NLP. It involves determining the sentiment or emotional tone expressed in a piece of text, such as positive, negative, or neutral. Sentiment analysis algorithms use various techniques, including lexicon-based approaches and machine learning models, to classify and quantify sentiment in textual data. This technology is widely used to analyze customer feedback, social media posts, and online reviews.

Question answering systems are designed to understand and respond to natural language queries. These systems employ techniques like information retrieval, knowledge representation, and natural language understanding to process user queries and provide relevant answers. Examples include virtual assistants like Siri, Alexa, and Google Assistant, which can respond to user questions and perform tasks based on the understanding of the query.

Text generation is another fascinating application of NLP. It involves the creation of human-like text by AI systems. Language generation models, such as recurrent neural networks (RNNs) and transformers, learn patterns from large textual datasets and generate coherent and contextually relevant sentences. Examples include chatbots, automated article writing, and language models like OpenAI's GPT.

NLP techniques rely on a combination of linguistic knowledge, machine learning algorithms, and large datasets to achieve accurate and meaningful language processing. They involve tasks like part-of-speech tagging, syntactic parsing, named entity recognition, and semantic analysis, which enable machines to understand the structure, meaning, and context of language.

NLP has seen significant advancements in recent years, thanks to the availability of large, annotated datasets, the development of deep learning models, and the progress in computational power. These advancements have revolutionized the field, enabling machines to understand and generate language with increasing accuracy and sophistication.

In summary, Natural Language Processing (NLP) is a vital building block of AI that focuses on enabling computers to understand, interpret, and generate human language. Through techniques like speech recognition, language translation, sentiment analysis, question answering, and text generation, NLP enables machines to process and interact with language in ways that resemble human understanding. This has wide-ranging applications in areas such as virtual assistants, machine translation, sentiment analysis, and automated content generation.

VII. Computer Vision

Computer Vision is a field of artificial intelligence (AI) that focuses on enabling machines to analyze, interpret, and understand visual information from images or videos. It allows AI systems to see and make sense of the visual world, similar to how humans perceive and interpret visual stimuli.

Computer Vision involves a range of tasks and techniques that aim to extract meaningful information from visual data. These tasks include object detection, image classification, facial recognition, scene understanding, and image generation. By analyzing visual content, AI systems can make decisions, provide insights, and interact with the world in a more visual and intuitive manner.

Object detection is one of the fundamental tasks in Computer Vision. It involves locating and identifying objects within an image or video. Object detection algorithms analyze visual data and draw bounding boxes around the objects of interest, enabling AI systems to recognize and differentiate between various objects. For example, object detection is used in self-driving cars to detect pedestrians, traffic signs, and other vehicles on the road.

Image classification is another essential task in Computer Vision. It involves assigning labels or categories to images based on their content. Image classification algorithms learn from a dataset of labeled images and can accurately classify new, unseen images into predefined categories. For example, image classification is used in medical imaging to classify X-ray images as normal or abnormal.

Facial recognition is a specialized application of Computer Vision that focuses on identifying and verifying individuals based on their facial features. Facial recognition algorithms analyze facial landmarks, such as the position of the eyes, nose, and mouth, and compare them with a database of known faces. This technology is used for identity verification, access control, and security systems.

Scene understanding is a task that involves extracting higher-level information and context from visual scenes. It goes beyond simple object detection and classification by considering the relationships and interactions between objects within a scene. Scene understanding algorithms can infer the spatial layout, object relationships, and scene semantics, enabling AI systems to understand complex visual scenes and make informed decisions.

Image generation is an exciting application of Computer Vision where AI systems are trained to generate new images or modify existing ones. Generative models, such as generative adversarial networks (GANs) and variational autoencoders

(VAEs), learn from a dataset of images and generate new images that resemble the training data. This technology has applications in areas like creative design, virtual reality, and entertainment.

Computer Vision techniques rely on a combination of image processing, pattern recognition, and machine learning algorithms. They analyze visual features such as colors, textures, shapes, and spatial relationships to extract meaningful information from images and videos. These algorithms learn from large, labeled datasets and can recognize patterns and objects in visual data with remarkable accuracy.

Computer Vision has numerous real-world applications across various industries. It is used in autonomous vehicles for object detection and pedestrian recognition, in surveillance systems for security and monitoring, in medical imaging for diagnosis and analysis, in robotics for object manipulation, and in augmented reality for virtual object placement and interaction.

In summary, Computer Vision is a crucial building block of AI that enables machines to analyze and understand visual information from images or videos. Through tasks like object detection, image classification, facial recognition, scene understanding, and image generation, Computer Vision allows AI systems to perceive and interpret the visual world. This has significant applications in areas such as autonomous vehicles, surveillance, healthcare, robotics, and augmented reality.

VIII.   Robotics

Robotics is a field that combines artificial intelligence (AI) with physical machines, known as robots, to create intelligent systems capable of sensing, acting, and interacting with the physical world. It involves designing and developing robots that can perceive their environment, make decisions based on that perception, and control their movements to perform specific tasks.

One of the key aspects of robotics is perception, which refers to a robot's ability to sense and understand its surroundings. Robots use various sensors, such as cameras, lidar, and proximity sensors, to gather data about their environment. By analyzing this sensory information, AI algorithms can extract meaningful features and understand the state of the world around them. For example, a robot equipped with a camera can use computer vision techniques to recognize objects, navigate through obstacles, and interact with its environment.

Decision-making is another crucial component of robotics. AI algorithms enable robots to analyze the information gathered from sensors, reason about it, and make intelligent decisions or take appropriate actions. These algorithms can utilize techniques like machine learning, planning, and control theory to process the sensory data, evaluate different options, and determine the best course of action. For instance, a robot in a manufacturing setting can use AI algorithms to optimize its movements and perform tasks efficiently.

Motor control is the aspect of robotics that enables robots to physically interact with the environment and manipulate objects. AI algorithms can control the motors and actuators of a robot, allowing it to move, grasp objects, and perform complex actions. By integrating AI with motor control, robots can achieve precise and coordinated movements, mimicking human-like dexterity. For example, a robot arm in a warehouse can use AI algorithms to grasp and stack objects efficiently.

Robotic systems are used in a wide range of applications and domains. In industrial settings, robots are used for tasks such as assembly, welding, and packaging, where they can perform repetitive and labor-intensive tasks with precision and speed. In healthcare, robots can assist in surgeries, rehabilitation, and patient care. In agriculture, robots can automate farming tasks, such as planting and harvesting. In exploration, robots are used in space missions and deep-sea exploration, where they can go to hazardous environments that are inaccessible to humans.

The field of robotics continues to advance rapidly, driven by advancements in AI and other technologies. Collaborative robots, known as cobots, are designed to work alongside humans, enhancing productivity and safety in manufacturing environments. Swarm robotics explores the coordination of multiple robots to accomplish complex tasks collectively. Autonomous drones are an example of robots that can navigate and make decisions in dynamic environments.

In summary, robotics combines AI and physical machines to create intelligent systems that can sense, act, and interact with the physical world. By incorporating perception, decision-making, and motor control, robots can perform tasks in various domains, ranging from manufacturing and healthcare to agriculture and exploration. Robotics play a vital role in improving efficiency, safety, and productivity across different industries, and its advancements continue to push the boundaries of what robots can achieve.

IX. Knowledge Representation

Knowledge Representation is a fundamental aspect of artificial intelligence (AI) that focuses on organizing and structuring information in a format that AI systems can understand and use for reasoning and decision-making. It involves representing knowledge and facts in a way that allows AI algorithms to make sense of the information and draw meaningful conclusions from it.

The goal of Knowledge Representation is to capture and organize knowledge in a structured form that can be processed by AI systems. This structured representation enables AI algorithms to reason, infer new information, and make intelligent decisions based on the available knowledge.

One common technique used in Knowledge Representation is ontologies. An ontology is a formal representation of knowledge that defines concepts, relationships, and properties within a specific domain. It provides a standardized vocabulary and set of rules that enable AI systems to understand and reason about the domain. For example, in medical ontology, concepts like "disease," "symptom," and "treatment" would be defined, along with their relationships and properties.

Semantic networks are another technique used in Knowledge Representation. A semantic network represents knowledge as a network of interconnected nodes, where each node represents a concept or entity, and the links between nodes represent relationships between them. For instance, in a semantic network for a restaurant domain, nodes could represent entities like "restaurant," "cuisine," and "location," and the links between them would define relationships such as "serves" or "located in."

Knowledge graphs are another powerful tool for representing knowledge. A knowledge graph is a graph-based structure that represents knowledge as entities, properties, and relationships between them. It allows for the integration of information from multiple sources and provides a rich representation of knowledge that can be used for reasoning and inference. For example, a knowledge graph could represent entities like "person," "organization," and "location," along with relationships like "works for" or "born in."

By utilizing techniques such as ontologies, semantic networks, and knowledge graphs, AI systems can represent knowledge in a way that enables reasoning and decision-making. For example, an AI system with medical ontology could reason about a patient's symptoms, medical history, and treatment options to make accurate diagnoses or treatment recommendations.

Knowledge Representation is crucial in AI because it allows systems to move beyond simple data processing and access a deeper understanding of the world. It enables AI algorithms to leverage prior knowledge and apply it to new situations, making more informed decisions. The structured representation of knowledge also facilitates the integration of information from different sources, enabling AI systems to make connections and discover new insights.

In summary, Knowledge Representation is a vital building block of AI that involves organizing and structuring information in a format that AI systems can understand and use for reasoning and decision-making. Techniques like ontologies, semantic networks, and knowledge graphs provide structured representations of knowledge, allowing AI algorithms to reason, infer new information, and make intelligent decisions. Knowledge Representation enhances the capabilities of AI systems, enabling them to leverage prior knowledge, integrate information, and gain a deeper understanding of the world.

X. Reinforcement Learning

Reinforcement Learning is a type of machine learning that enables AI agents to learn how to make decisions through trial-and-error interactions with an environment. It is inspired by the way humans and animals learn through positive rewards and negative consequences. Reinforcement Learning allows AI agents to learn optimal behaviors by maximizing rewards and minimizing penalties.

In Reinforcement Learning, an AI agent interacts with an environment and takes actions to achieve a specific goal. After each action, the agent receives feedback in the form of rewards or penalties. The rewards indicate the desirability of the agent's action, while penalties represent undesirable outcomes. The agent's objective is to learn a policy or strategy that maximizes the cumulative reward over time.

The learning process in Reinforcement Learning can be understood through the concept of an "agent-environment loop." The agent observes the current state of the environment, selects an action based on its learned policy, and then receives feedback in the form of a reward. The agent uses this feedback to update its knowledge and improve its decision-making abilities. This iterative process continues until the agent learns the optimal behavior.

One of the key components of Reinforcement Learning is the reward function. The reward function assigns a numeric value to each state-action pair, indicating the desirability of that action in that particular state. The agent's goal is to learn a policy that maximizes the cumulative reward over time. For example, in a game of chess, the agent receives a positive reward for winning the game, a negative reward for losing, and intermediate rewards for making good moves or avoiding bad moves.

Reinforcement Learning algorithms employ various techniques to learn optimal policies. One common approach is Q-learning, which uses a Q-table to store and update the expected rewards for each state-action pair. Through repeated interactions with the environment, the agent updates the Q-table, learning the optimal actions to take in each state. Another popular technique is the use of deep neural networks in combination with Reinforcement Learning, known as Deep Reinforcement Learning. This approach enables AI agents to learn directly from raw sensory inputs, such as images or sensor data.

Reinforcement Learning has been successfully applied in various domains. For example, in robotics, Reinforcement Learning can be used to teach robots how to perform complex tasks by interacting with the environment and learning from feedback. In autonomous driving, Reinforcement Learning can enable self-driving cars to learn optimal navigation strategies. In game playing, Reinforcement Learning has been used to train AI agents to outperform human experts in games like Go, chess, and poker.

Reinforcement Learning is a powerful building block of AI because it allows agents to learn optimal behaviors without explicit programming. By continuously interacting with the environment and receiving feedback, AI agents can adapt their decision-making strategies and improve their performance over time. This approach is particularly useful in situations where there is no readily available training data or explicit rules to follow.

In summary, Reinforcement Learning is a building block of AI that enables agents to learn optimal behaviors through trial-and-error interactions with an environment. By receiving rewards or penalties based on their actions, AI agents can optimize their decision-making processes over time. Reinforcement Learning algorithms employ techniques such as Q-learning and Deep Reinforcement Learning to learn policies that maximize cumulative rewards. This approach has been successfully applied in various domains, including robotics, autonomous driving, and game playing.

XI. Expert Systems

Expert Systems are AI systems designed to mimic the knowledge and decision-making abilities of human experts in specific domains. They are built using rule-based reasoning and knowledge bases to provide expert-level advice or make informed decisions. Expert Systems aim to capture and emulate the expertise and problem-solving strategies of human experts, enabling them to assist users in solving complex problems.

At the heart of an Expert System is its knowledge base, which contains a collection of rules and facts about a particular domain. These rules represent the expertise and decision-making logic of human experts. Each rule consists of a condition and an action. The condition specifies the set of conditions that must be met, while the action describes the corresponding action or recommendation to be taken.

For example, in a medical Expert System, a rule might state: "If the patient has a fever, cough, and difficulty breathing, then recommend testing for respiratory infection." The Expert System would use these rules to analyze a patient's symptoms and provide recommendations based on the matching rules and facts in its knowledge base.

Expert Systems also incorporates an inference engine, which is responsible for applying the rules to the available data and making logical inferences. The inference engine follows a set of reasoning mechanisms to evaluate the conditions in the rules and determine the appropriate actions or recommendations.

One common reasoning mechanism used in Expert Systems is forward chaining. It starts with the available facts and applies rules that match those facts. As new facts are generated, the system continues to apply relevant rules until a goal or solution is reached. For example, in a diagnostic Expert System, the system might start with symptoms reported by the patient and apply rules to identify potential diseases.

Another reasoning mechanism used in Expert Systems is backward chaining. It starts with a goal or desired outcome and works backward to find the set of conditions that need to be satisfied to achieve that goal. For instance, in a financial planning Expert System, the system might start with a goal of retirement planning and use backward chaining to determine the necessary financial steps and strategies.

Expert Systems have been applied in various domains, including medicine, finance, engineering, and troubleshooting. They have proven particularly valuable in domains where access to human experts is limited, expensive, or time-consuming. Expert Systems can provide consistent and reliable advice based on the accumulated knowledge and experience of human experts.

However, it is important to note that Expert Systems have limitations. They rely on the accuracy and completeness of the knowledge base and may struggle with handling uncertain or ambiguous information. Additionally, they typically require manual encoding of expert knowledge, which can be time-consuming and challenging to update as new knowledge emerges.

In summary, Expert Systems are a building block of AI that mimic the knowledge and decision-making abilities of human experts in specific domains. They use rule-based reasoning and knowledge bases to provide expert-level advice or make informed decisions. Expert Systems rely on a collection of rules and facts to emulate human expertise, and their inference engines apply reasoning mechanisms to evaluate conditions and generate recommendations. They have been applied in various domains and provide valuable assistance where access to human experts is limited or costly.

XII.   Data Mining

Data Mining is a crucial building block of AI that refers to the process of discovering patterns, correlations, and insights from large datasets. It involves using computational techniques to extract valuable knowledge from data, enabling organizations and researchers to make informed decisions and predictions.

In Data Mining, the goal is to uncover hidden patterns and relationships that may not be apparent at first glance. It goes beyond simple data analysis by applying advanced algorithms and statistical methods to identify meaningful information and extract actionable insights. Data Mining allows us to delve into vast amounts of data and extract knowledge that can be used to solve problems, optimize processes, and gain a competitive advantage.

One of the fundamental techniques in Data Mining is clustering. Clustering algorithms group similar data points together based on their characteristics or attributes. For example, in customer segmentation, clustering can be used to identify distinct groups of customers based on their purchasing behavior, demographics, or preferences. This information can then be used for targeted marketing or personalized recommendations.

Classification is another important technique in Data Mining. It involves building models that can classify data into predefined categories or classes. For instance, in email spam detection, a classification model can be trained to distinguish between spam and legitimate emails based on various features like the subject, sender, and content. This allows for automatic filtering and sorting of incoming emails.

Association rule mining is another valuable technique in Data Mining. It aims to discover relationships and associations between items in a dataset. For example, in a retail setting, association rule mining can uncover interesting patterns like "Customers who buy diapers are likely to also purchase baby formula." These associations can be used for cross-selling and targeted marketing strategies.

Data Mining also involves techniques such as regression analysis, anomaly detection, and sequential pattern mining, depending on the specific objectives and characteristics of the dataset.

The process of Data Mining typically involves several steps, including data preprocessing, where the data is cleaned, transformed, and prepared for analysis; exploratory data analysis, which involves visualizing and understanding the data; model building, where algorithms are applied to extract patterns and insights; and evaluation, where the performance and validity of the models are assessed.

Data Mining has numerous practical applications across industries. For example, in healthcare, Data Mining can be used to analyze patient data to identify disease risk factors and improve treatment outcomes. In finance, it can help detect fraudulent transactions and predict market trends. In manufacturing, Data Mining can optimize production processes and reduce waste. The possibilities are vast.

In summary, Data Mining is a fundamental building block of AI that allows us to uncover patterns, correlations, and insights from large datasets. By applying clustering, classification, association rule mining, and other techniques, Data Mining enables us to extract valuable knowledge and make informed decisions. It has wide-ranging applications in various domains, empowering organizations to gain insights, optimize processes, and enhance decision-making.

XIII.   Cognitive Computing

Cognitive Computing is a significant building block of AI that aims to simulate human-like intelligence by combining various AI techniques with an understanding of human cognition. It focuses on creating AI systems that can reason, learn, and interact with humans in a more human-like manner, bridging the gap between machines and humans.

Cognitive Computing draws inspiration from how the human brain works, seeking to replicate cognitive processes such as perception, understanding, learning, and decision-making. By incorporating techniques from fields like natural language processing, pattern recognition, machine learning, and knowledge representation, Cognitive Computing enables AI systems to interpret and respond to human inputs in a more intelligent and context-aware manner.

One key aspect of Cognitive Computing is natural language processing (NLP). NLP enables AI systems to understand and interpret human language, including speech and text. By processing and analyzing linguistic data, AI systems can extract meaning, infer intent, and engage in meaningful conversations with humans. For example, virtual assistants like Siri and Alexa utilize NLP techniques to understand voice commands and provide relevant responses.

Pattern recognition is another crucial component of Cognitive Computing. AI systems are trained to recognize patterns in data, such as images, sounds, or text, to extract meaningful information and make informed decisions. This ability to recognize patterns allows AI systems to identify objects, detect anomalies, or predict future outcomes. For instance, image recognition technology can identify objects in photos or videos, enabling applications like facial recognition or object detection.

Machine learning is a fundamental technique within Cognitive Computing. It involves training AI systems on large datasets to learn patterns and relationships, enabling them to make predictions or take actions based on new inputs. By learning from examples and feedback, machine learning algorithms can improve their performance over time. For instance, recommender systems in e-commerce platforms use machine learning to analyze user preferences and make personalized product recommendations.

Knowledge representation is another important aspect of Cognitive Computing. It involves organizing and structuring information in a format that AI systems can understand and use for reasoning and decision-making. By representing knowledge in a structured manner, AI systems can draw upon existing knowledge to answer questions, solve problems, and make informed decisions. Knowledge graphs and ontologies are commonly used techniques for representing knowledge.

Cognitive Computing has wide-ranging applications in areas such as healthcare, finance, customer service, and education. For example, in healthcare, Cognitive Computing can help analyze patient data to support diagnosis and treatment decisions. In finance, it can assist in analyzing market trends and making investment recommendations. In customer service, it can enable chatbots to understand and respond to customer inquiries. These are just a few examples of how Cognitive Computing is revolutionizing various industries.

In summary, Cognitive Computing is a crucial building block of AI that combines techniques from natural language processing, pattern recognition, machine learning, and knowledge representation to create AI systems that simulate human-like intelligence. By focusing on reasoning, learning, and interaction, Cognitive Computing aims to bridge the gap between machines and humans, enabling AI systems to understand and respond to human inputs in a more intelligent and context-aware manner.

These building blocks work in conjunction to create AI systems with various capabilities and functionalities. The specific combination and implementation of these components depend on the AI application and the problem being addressed. By understanding and harnessing these building blocks, AI researchers and developers can create powerful and intelligent systems that have the potential to revolutionize industries and enhance our lives.

3

"AI is not about building machines that are more intelligent than humans; it's about building machines that can help humans be more intelligent."

- Erik Brynjolfsson, Economist and AI Researcher

The Rise of AI:
A Historical Perspective

Early Endeavors in AI: From Turing to the AI Winter

The journey of Artificial Intelligence (AI) from an intriguing concept to a real-world technology has been marked by incredible innovations, dashed expectations, and eventual resurgence. To understand how we've arrived at our current state of AI, let's look back at the early endeavors in this field.

I. The Turing Era (1940s-1950s)

The Turing Era, which took place in the 1940s and 1950s, marked a significant period in the advancement of artificial intelligence (AI) and computing. It was during this time that the renowned British mathematician and computer scientist Alan Turing made groundbreaking contributions that laid the foundation for AI research.

During World War II, Turing worked at the Government Code and Cypher School in the United Kingdom, where he played a crucial role in decrypting German Enigma machine messages, contributing significantly to the Allied war effort. His work on breaking the Enigma code demonstrated the power of computation in solving complex problems and provided insights into the potential of machines to perform intelligent tasks.

One of Turing's most notable contributions was the development of the concept of a universal computing machine, now known as the Turing machine. The Turing machine was a theoretical device that laid the groundwork for modern computer architecture. It introduced the concept of a programmable machine capable of executing any algorithm, thereby establishing the notion of computation as a universal concept. Turing's work on the Turing machine provided a theoretical framework for understanding the limits and possibilities of computation, which later influenced the field of AI.

In his seminal paper titled "Computing Machinery and Intelligence" published in 1950, Turing proposed the famous "Turing Test" as a measure of machine intelligence. The Turing Test involves a human evaluator engaging in a conversation with both a human and a machine. If the evaluator is unable to distinguish between the two based on their responses, the machine is said to exhibit intelligent behavior. The Turing Test became a significant milestone in AI research and continues to be a benchmark for evaluating machine intelligence.

Turing's ideas and concepts formed the basis for early AI research and influenced the development of AI systems and algorithms. His work inspired researchers to explore the possibility of creating intelligent machines capable of performing tasks that were traditionally considered to require human intelligence.

One notable example of AI advancement during The Turing Era is the Logic Theorist, developed by Allen Newell and Herbert A. Simon in 1955. The Logic Theorist was an early AI program that aimed to prove mathematical theorems using symbolic logic. It demonstrated the potential of AI systems to mimic human problem-solving and reasoning.

The Turing Era was a critical period that shaped the direction of AI research. Turing's concepts, such as the universal computing machine and the Turing Test, laid the groundwork for the development of AI systems and algorithms. His ideas inspired subsequent generations of researchers to explore the capabilities of machines and their potential to exhibit intelligent behavior.

II.   Early Optimism and Achievements (1950s-1960s)

The period of the 1950s to the 1960s was marked by great optimism and significant achievements in the field of artificial intelligence (AI). Researchers and pioneers in AI believed that intelligent machines could be created within a relatively short period of time.

During this era, AI researchers made remarkable progress in developing new algorithms, designing intelligent systems, and exploring the potential applications of AI. The optimism stemmed from the belief that the fundamental principles of intelligence could be understood and replicated in machines.

One of the notable achievements during this period was the development of the Logic Theorist by Allen Newell and Herbert A. Simon in 1955. The Logic Theorist was an AI program capable of proving mathematical theorems using symbolic logic.

Its success in proving complex theorems demonstrated the potential of AI systems to mimic human reasoning and problem-solving.

Another significant milestone was the development of the General Problem Solver (GPS) by Newell, Simon, and J.C. Shaw in 1957. The GPS was a problem-solving program that could generate solutions to a wide range of problems by using a set of general rules. It showcased the power of AI systems in tackling complex tasks and demonstrated the potential of AI to emulate human-like problem-solving abilities.

Furthermore, in 1956, the Dartmouth Conference marked the birth of AI as a field of research. The conference brought together leading researchers including John McCarthy, Marvin Minsky, and Claude Shannon, who shared their ideas and set the agenda for AI research. The discussions and collaborations that took place at the Dartmouth Conference contributed to the rapid advancement of AI during this period.

AI researchers during this time were also optimistic about the potential of machine learning. In 1956, Arthur Samuel developed a program that could play checkers and improve its performance through self-learning. This breakthrough demonstrated that machines could acquire knowledge and improve their performance through experience, a concept fundamental to machine learning.

The achievements and progress made during this early era fueled high expectations for the future of AI. Researchers believed that within a few decades, machines would possess human-like intelligence and be capable of performing tasks across various domains.

However, as the field progressed, challenges and limitations began to emerge, leading to what was known as the "AI Winter" in the late 1960s and 1970s. Nevertheless, the achievements of this early era of AI laid the groundwork for subsequent advancements in AI. The pioneering work on symbolic logic, problem-solving, and machine learning set the stage for further developments and led to the growth and expansion of the field.

III. AI Winter (1970s-1990s)

The period from the 1970s to the 1990s is often referred to as the AI Winter—a time of reduced funding, limited progress, and decreased enthusiasm for artificial intelligence (AI) research. The AI Winter was characterized by a significant decline in interest and support for AI as initial promises and expectations were not met.

During this era, several factors contributed to the decline of AI research. One major factor was the inability of AI systems to live up to the lofty expectations set in the early years. AI was initially believed to be capable of human-level intelligence and solving complex problems across various domains. However, as researchers delved deeper into AI, they encountered challenges and limitations that proved more difficult to overcome than initially anticipated.

One contributing factor to the AI Winter was the high computational and resource requirements of early AI systems. Many AI algorithms and approaches required substantial computational power and memory, which was not readily available at the time. As a result, progress in AI research was hindered by the limitations of existing hardware and computing capabilities.

Another factor was the lack of significant breakthroughs in AI during this period. Despite efforts to advance AI technologies, researchers struggled to overcome challenges such as knowledge representation, natural language processing, and real-world reasoning. As a result, AI systems often fell short of achieving human-level performance, leading to disappointment and waning interest.

Additionally, funding for AI research experienced a decline during the AI Winter. The failure to meet early expectations led to skepticism among funding agencies and investors, causing a decrease in financial support for AI projects. The reduction in funding limited the resources available for research and development, further slowing down progress in the field.

During the AI Winter, interest in AI shifted towards more practical applications and specialized domains, such as expert systems and rule-based systems. These systems focused on narrow tasks and specific domains rather than attempting to replicate human-like intelligence in a broad sense. Although these applications provided some value, they did not meet the grand ambitions of early AI research.

However, despite the challenges and setbacks of the AI Winter, important advancements and foundational work were still accomplished. Researchers made progress in areas such as expert systems, natural language processing, and knowledge-based systems, laying the groundwork for future developments.

The AI Winter eventually ended in the late 1990s when new breakthroughs and advancements reignited interest and enthusiasm for AI. Progress in areas like machine learning, data availability, and computational power led to the emergence of new approaches and techniques that revitalized the field.

The AI Winter serves as a valuable lesson in managing expectations and understanding the complexities of AI research. It highlighted the need for a more realistic and incremental approach to AI development, focusing on specific tasks and domains before attempting to achieve general artificial intelligence.

IV. The Resurgence of AI: Internet Era and the Role of Big Data

The advent of the Internet and the exponential growth of digital data have been instrumental in fueling the resurgence of Artificial Intelligence (AI) in the 21st century. This section will explore the connection between the Internet, Big Data, and the modern development of AI, highlighting the transformative impact they have had on the field.

The Internet: A Data Explosion

With the widespread adoption of the Internet, the world has witnessed an unprecedented explosion of data. Today, billions of people use the Internet, generating a wealth of digital information in the form of text, images, videos, and more. By 2021, it was estimated that 2.5 quintillion bytes of data were being created every day (Marr, 2018).

Big Data: Feeding AI's Hunger

The abundance of digital data, known as Big Data, has provided the fuel that AI systems need to learn and grow. Machine learning algorithms, the driving force behind modern AI, rely on vast amounts of data to identify patterns and make predictions. As Andrew Ng, a leading AI researcher, explains, "The rocket engine is the deep learning models, and the fuel is the huge amounts of data we can feed to these algorithms" (Ng, 2016).

The Role of Big Data in AI Advancements

The availability of Big Data has paved the way for advancements in AI, such as:

Natural Language Processing (NLP): The Internet has produced vast amounts of text data, enabling AI systems to analyze and understand human languages more effectively (Mikolov et al., 2013).

Image Recognition: The millions of images available online have provided ample training data for AI algorithms to recognize objects and patterns in images (Krizhevsky et al., 2012).

Recommendation Systems: Online platforms like Amazon and Netflix use AI algorithms to analyze user data and make personalized recommendations based on individual preferences and behavior (Koren et al., 2009).

Challenges and Considerations

While Big Data has accelerated AI development, it also raises concerns, such as data privacy, security, and potential biases in the data. As AI systems increasingly rely on large datasets, it is essential to address these issues and ensure that AI algorithms provide fair and accurate outcomes.

The Internet era and the rise of Big Data have been instrumental in reviving and advancing AI research. As we continue to generate more data and refine AI algorithms, the capabilities of AI will continue to expand, shaping the future of technology and society.

Milestones in AI Development: Chess, Go, and Beyond

The journey of Artificial Intelligence (AI) has been marked by several key milestones that have demonstrated the potential of this transformative technology. These milestones, achieved in the domains of games like Chess and Go, have dramatically shown the world what AI can do.

I. Deep Blue and the Chess Victory (1997)

In 1997, IBM's Deep Blue made headlines worldwide when it defeated world chess champion Garry Kasparov in a six-game match. This event was a significant moment in the history of AI because chess is a game that requires strategic thinking and forward planning—attributes that were long thought to be unique to human intelligence.

Deep Blue's victory showcased the potential of AI in complex problem-solving tasks. The machine used brute force computation, examining 200 million possible moves per second, and advanced evaluation functions to determine the desirability of a position (Campbell, Hoane Jr, & Hsu, 2002).

II. AlphaGo and the Game of Go (2016)

While Deep Blue's victory was impressive, another milestone in 2016 underscored the progress in AI. This time, the game was Go, an ancient Chinese board game considered much more complex than chess due to the vast number of potential moves.

DeepMind's AlphaGo defeated world champion Lee Sedol in a five-game match, a feat that was thought to be at least a decade away. Unlike Deep Blue, which relied on brute force computation, AlphaGo used deep learning and reinforcement learning techniques to train itself by playing millions of games against itself and improving over time (Silver et al., 2016).

III. OpenAI and Dota 2 (2018)

Pushing the boundaries even further, in 2018, OpenAI's AI system, OpenAI Five, competed against professional human teams at Dota 2, a popular and complex online video game. OpenAI Five demonstrated the ability to handle the game's intricate dynamics and unpredictable gameplay, illustrating AI's potential in real-time decision making and team-based strategy games (OpenAI, 2018).

IV. The Significance and Beyond

These milestones have dramatically shown AI's potential, especially in problem-solving and strategic thinking. However, it's essential to understand that these achievements, while impressive, still fall under the category of Narrow AI – AI designed to perform a specific task. The complexity and adaptability of human intelligence, covering a wide range of tasks and environments, remain challenges for AI research.

4

"We are at the beginning of a golden age of AI."

- Jensen Huang, CEO of NVIDIA (NVIDIA Blog, 2018).

AI Today: Applications and Innovations

Industry

I. The Rise of Intelligent Automation

Automation, in its fundamental essence, is the use of technology to perform tasks that once required human effort. With the advent of artificial intelligence (AI), the potential of automation has seen a profound expansion. The incorporation of AI into automation has allowed it to cross the boundaries from simply executing physical tasks to performing cognitive ones. This evolution is recognized as intelligent automation.

In Manufacturing

The manufacturing industry, a bedrock of any economy, is currently undergoing a transformative change thanks to AI. Intelligent automation is transforming how we manufacture goods, streamlining production processes, enhancing quality control, and making supply chain logistics more efficient.

AI can analyze multiple variables in production processes to find the optimal production parameters, leading to a reduction in production waste and costs. Quality control is improved as AI can detect minute inconsistencies or faults that might be missed by human inspectors.

Safety in manufacturing has been significantly enhanced through the use of autonomous robots. These AI-powered robots are capable of performing tasks in environments that are deemed hazardous for humans. These tasks may involve handling dangerous substances or operating in high-temperature conditions. This not only increases worker safety but also leads to an increase in overall efficiency, as these robots can work round-the-clock without breaks (Banga, 2018).

In Customer Service

Customer service is another area where AI-driven automation has made a substantial impact. AI chatbots and virtual assistants are increasingly being deployed to handle routine inquiries, thus providing 24/7 customer support. This shift not only provides customers with instant support but also liberates human agents to handle more complex and sensitive customer interactions.

An AI chatbot can assist with basic inquiries such as tracking an order or providing information about services or operating hours. If a customer issue necessitates human intervention, the chatbot can seamlessly transfer the interaction to a human representative. This blend of AI and human interaction can enhance customer satisfaction while making customer service operations more efficient (Xu, Liu, & Li, 2021).

As AI technology advances, the scope and effectiveness of intelligent automation will continue to grow. The ongoing challenge is to harness these capabilities responsibly, ensuring they serve to augment human capabilities and facilitate greater efficiencies, rather than displace human roles in a disruptive manner.

II. AI in Industry: Predictive Analysis

Predictive analysis is a powerful AI-driven tool that is significantly influencing various industries. In essence, it involves using data, machine learning algorithms, and statistical techniques to forecast future events and trends. By harnessing the power of AI, predictive analysis can analyze vast amounts of historical data, identify patterns, and make accurate predictions that guide decision-making. This section delves into the role of predictive analysis in various industries, making it accessible to the average reader.

Healthcare

AI-enabled predictive analysis is transforming the healthcare industry by offering insights into patient care and treatment outcomes. Machine learning algorithms can analyze Electronic Health Records (EHRs), medical imaging, and patient data to identify patterns that can help healthcare providers make more informed decisions regarding diagnoses and treatment plans (Raghupathi & Raghupathi, 2014).

For example, predictive analysis can be employed to identify patients at a higher risk of readmission or complications after a specific procedure. This information enables doctors to take preventive measures or adjust treatments to improve patient outcomes (Bayati et al., 2014).

Finance

The finance industry has embraced AI and predictive analysis to make more accurate forecasts about market trends, optimize trading strategies, and assess credit risks. By analyzing historical financial data, AI can identify patterns and correlations that can help investors make informed decisions about their investments (Patel et al., 2015).

AI-driven predictive analysis has also enabled banks and financial institutions to assess the creditworthiness of individuals and businesses more accurately. This ensures that financial services are offered more effectively while minimizing the risks associated with lending (Jagtiani & Lemieux, 2017).

Retail

In the retail sector, predictive analysis helps businesses to forecast customer demands, manage inventory, and personalize marketing efforts. Retailers can use AI to analyze past sales data and customer behavior patterns to predict which products are likely to sell well in the future. This information allows retailers to make data-driven decisions about stocking inventory, pricing strategies, and promotional campaigns (Chen et al., 2012).

Moreover, AI-powered predictive analysis can identify individual customer preferences, helping retailers to create personalized marketing content that appeals to their target audience, thereby improving customer engagement and loyalty (Li & Karahanna, 2015).

AI-driven predictive analysis has become an invaluable tool across various industries, enhancing decision-making, improving efficiency, and promoting innovation. As AI technology continues to evolve, its applications in predictive analysis will only grow, further impacting the way businesses operate and consumers experience products and services.

III. AI in Industry: Personalization

Personalization, driven by Artificial Intelligence, is revolutionizing the way businesses interact with their customers. AI systems can analyze vast amounts of data about users' behavior, preferences, and habits, using this information to offer tailored services and experiences. This section aims to explore the concept of personalization in AI and provide examples from various industries, all in layperson's terms.

E-Commerce

In the e-commerce industry, personalization manifests as AI recommending products based on the user's previous browsing and purchasing behavior. For example, if you've been searching for running shoes on an e-commerce platform, you might start seeing suggestions for similar products or related items like fitness apparel. Such recommendations are generated by machine learning algorithms, which analyze your past actions and predict what you might be interested in (Li & Karahanna, 2015).

Streaming Services

Similarly, streaming platforms like Netflix and Spotify use AI-driven personalization to recommend movies, series, or songs that you might enjoy. Their recommendation algorithms analyze your viewing or listening history, ratings, and even the behavior of other users with similar tastes to predict what content you might find engaging (Gomez-Uribe & Hunt, 2016).

Marketing

In the marketing field, businesses use AI to deliver personalized advertisements and promotional content. By analyzing your online activity, businesses can tailor their marketing messages to your interests, increasing the likelihood that you'll engage with their ads. This targeted approach is not only more effective but also leads to a better user experience, as customers receive ads that are relevant to their needs (Huang & Rust, 2018).

Education

AI-driven personalization also extends to the education sector, where it's known as adaptive learning. Adaptive learning platforms analyze a student's performance to customize their learning experience. By identifying a student's strengths and weaknesses, these platforms can adjust the pace of instruction, offer tailored practice problems, provide personalized feedback, enhancing learning outcomes (Walkington, 2013).

AI-driven personalization is reshaping the customer experience across various industries. As AI technology continues to advance, it will provide even more opportunities for businesses to tailor their services and create unique, engaging experiences for their customers.

Healthcare

I. AI in Healthcare: Diagnosis

Artificial intelligence is dramatically altering healthcare, with one of its most prominent applications being disease diagnosis. Using advanced algorithms, AI can analyze medical data, spot patterns, and assist in diagnosing a wide range of conditions. This section will demystify how AI aids in diagnosis and provide relevant examples from the field.

Medical Imaging

Medical imaging, like X-rays, CT scans, or MRIs, is a critical tool for diagnosing many conditions. Traditionally, these images have been reviewed by radiologists who rely on their expertise and experience to identify abnormalities. Today, AI algorithms are being developed and employed to assist in this process. These algorithms are trained on vast datasets of medical images and corresponding diagnoses, learning to spot patterns or features indicative of specific conditions.

For example, researchers have developed AI systems capable of detecting lung cancer in CT scans, often achieving similar or even better performance than human radiologists. These algorithms can identify tiny nodules or growths that may be easily overlooked (Ardila et al., 2019).

Electronic Health Records

Electronic Health Records (EHRs) are digital versions of patients' medical histories, including doctor's notes, laboratory results, and past treatments. Analyzing this data manually to predict or diagnose health conditions can be time-consuming and challenging. However, AI algorithms can quickly process and analyze these records, uncovering patterns or correlations that might indicate a disease.

For example, Google's DeepMind developed an AI system that could predict kidney injury up to 48 hours before it happened by analyzing a wide range of data in patients' EHRs (Futoma, Simons, Panch, Doshi-Velez, & Celi, 2020).

Genetic Analysis

AI also has the potential to revolutionize the field of genetic testing and diagnosis. Machine learning algorithms can analyze genetic data to identify mutations or patterns that may be associated with specific diseases. For instance, AI is being used to diagnose rare genetic disorders in children by analyzing their facial features from photographs (Ferry et al., 2020).

AI is rapidly becoming an indispensable tool in disease diagnosis, offering the potential to enhance accuracy, efficiency, and predictability in healthcare. As research advances, we can expect AI to play an even more significant role in diagnosing diseases, predicting outcomes, and tailoring treatments to individual patients.

II. Drug Discovery

Developing new drugs is an incredibly complex and costly process that traditionally takes years, if not decades, and billions of dollars. This is where artificial intelligence can step in and revolutionize the process. AI has the potential to significantly expedite drug discovery, reduce development costs, and improve the likelihood of finding effective therapeutics. This section will provide an overview of how AI is transforming drug discovery.

Identification of Drug Targets

A crucial first step in drug development is identifying the right biological target - usually a protein or a gene associated with a disease. AI can aid this process by sifting through vast amounts of biological data, identifying patterns and correlations that may suggest potential targets. For example, machine learning algorithms have been used to analyze genetic data from patients with specific diseases, identifying genes that may be critical to the disease process and could serve as promising drug targets (Costello et al., 2019).

Screening of Drug Candidates

Once potential targets are identified, the next step is to find chemical compounds that can affect these targets in a beneficial way. Traditional methods of screening these potential drug candidates involve extensive lab testing that can take years. AI, specifically a type of machine learning known as deep learning, can simulate how different chemical compounds might interact with the target, significantly accelerating the screening process.

Insilico Medicine, a biotech firm, demonstrated the power of AI in this regard by using a deep learning model to design, synthesize, and validate a novel drug candidate in just 46 days, a process that traditionally could have taken years (Zhavoronkov et al., 2019).

Predicting Drug Side Effects and Interactions

AI can also be used to predict potential side effects and interactions of drug candidates, which is a significant concern in drug development. By analyzing vast databases of known drug side effects and interactions, AI models can predict potential issues with new drug candidates, potentially saving years of work and millions of dollars in the later stages of drug development.

AI holds substantial promise in transforming drug discovery. Its potential to analyze vast datasets quickly and accurately, to predict interactions and side effects, and to expedite the overall discovery process, could significantly reduce the time and cost of bringing new drugs to market.

III. Patient Care

Artificial Intelligence (AI) is not only revolutionizing diagnosis and drug discovery in healthcare, but it's also transforming the way patient care is delivered. From enhancing personalized treatment plans to monitoring patient health and improving hospital logistics, AI is increasingly becoming a vital tool in patient care.

Personalized Treatment Plans

Every individual is unique, and so is their response to treatment. AI can help create more personalized treatment plans that consider the patient's unique genetic makeup, lifestyle, and other factors. This process, known as precision medicine, is becoming more feasible with the help of machine learning algorithms that can analyze vast amounts of data from various sources like genomic sequencing, electronic health records, and wearable devices (Collins & Varmus, 2015).

Patient Monitoring and Predictive Analytics

AI-powered remote monitoring tools can collect a wealth of data from patients in real-time, such as heart rate, blood pressure, and glucose levels. Machine learning algorithms can then analyze this data to identify patterns and predict potential health risks, allowing for proactive care. Google's DeepMind Health is one example of how AI can predict patient deterioration by analyzing electronic health records (Rajkomar et al., 2018).

Improved Hospital Logistics

AI can also help improve hospital logistics, leading to more efficient patient care. For example, machine learning algorithms can help manage patient flow in emergency departments or optimize scheduling in operating rooms, leading to shorter wait times and improved patient satisfaction (Zhang et al., 2020).

Digital Health Assistants

AI-enabled virtual health assistants, such as chatbots and virtual nurses, can provide round-the-clock support to patients, answering queries, offering medication reminders, and providing health education. This not only enhances patient care but also reduces the burden on healthcare professionals.

AI has the potential to significantly improve patient care by enabling personalized treatment, enhancing patient monitoring, streamlining hospital logistics, and offering digital health assistance. As AI technologies continue to evolve, they will undoubtedly play an increasingly integral role in patient care.

Entertainment

I. Video Games

Artificial Intelligence (AI) has been deeply integrated into the realm of video games, significantly enhancing the gameplay experience. From designing non-player characters (NPCs) with complex behaviors to procedural content generation and personalizing player experiences, AI's influence is far-reaching.

Non-Player Characters and AI

In video games, NPCs, or characters controlled by the game itself rather than by players, play a crucial role. They help set the scene, drive the plot, and often serve as allies or opponents for the player. AI has greatly expanded the capabilities of NPCs. For instance, AI can help NPCs react more realistically to player actions, creating a more immersive and dynamic gaming experience.

Consider Rockstar Games' "Red Dead Redemption 2", a game highly praised for its AI-driven NPCs. Characters in the game go about their daily routines and react differently based on players' actions and the game's storyline, creating a world that feels alive (Schreier, 2018).

Procedural Content Generation

AI has also revolutionized the way content is generated in games. Traditional game design involves manually creating levels, maps, and scenarios, a process that can be time-consuming and limiting. AI algorithms, particularly those based on machine learning, can generate new content procedurally, creating endless variations and enhancing replayability.

Minecraft's "Infinite World" feature is an excellent example of procedural content generation. The game uses a pseudo-random number generator and a set of rules to create vast, unique landscapes for players to explore (O'Dell, 2011).

Personalizing Player Experience

Personalization is a key aspect of modern gaming. AI algorithms can adapt to a player's behavior, modifying game difficulty or tailoring content to the player's preferences. For instance, "Left 4 Dead", a game developed by Valve, uses an AI "Director" that changes the placement and intensity of enemy encounters based on player performance (Valve, 2008).

AI has become an essential tool for creating dynamic, engaging, and personalized experiences in video games. As AI technology continues to advance, its impact on the gaming industry is likely to grow even more significant.

II. Movies

Artificial Intelligence (AI) is leaving its mark on the film industry, aiding in areas from movie production to viewer experience. It's creating special effects, curating personalized content, and even predicting box-office successes.

Special Effects and Animation

AI has significantly impacted the creation of visual effects and animation in movies. High-quality CGI (computer-generated imagery) and special effects, once the domain of big-budget films, have become more accessible and efficient with AI technology.

Take the use of deep learning algorithms for automating rotoscoping - the labor-intensive process of frame-by-frame editing used for creating effects like realistic background removal or color grading. This automation accelerates the production process and allows for more intricate and lifelike effects (Winfield & Bishop, 2020).

In animation, AI has been used to create more natural movements for characters. An example is Disney's use of AI for their animated films to create more realistic and dynamic crowd scenes, reducing the manual workload of animators (Lino, Christie & Thalmann, 2014).

Personalized Content Curation

AI algorithms are also used by streaming platforms like Netflix and Amazon Prime to recommend movies tailored to a viewer's preferences. These platforms analyze viewing habits, ratings given, and even the time spent watching specific movies to provide personalized content recommendations (Gomez-Uribe & Hunt, 2015).

Predicting Box-Office Success

AI has also been utilized to predict the success of films at the box office. Companies like ScriptBook use natural language processing (NLP) and machine learning to analyze movie scripts and predict their financial success before they are produced (Van den Oord, Dieleman, & Schrauwen, 2013).

AI has integrated itself into the film industry, streamlining production, enhancing viewer experiences, and predicting success. As technology continues to advance, its influence on the film industry is poised to grow even further.

III. Music

The music industry, from creation to consumption, is undergoing a transformation due to the emergence of Artificial Intelligence (AI). It influences how music is composed, distributed, and even how we discover new songs. AI is expanding the possibilities of music in ways we've never seen before.

Composition and Production

AI has made its way into the music composition process, enabling the creation of new tunes, and assisting musicians in their creative process. Companies like OpenAI's MuseNet and

Jukin have developed AI models that can generate music spanning multiple genres and styles (Payne, 2019).

AI is also used in music mastering, the final step in audio post-production. AI-powered software like LANDR allows independent musicians to polish their tracks, balancing the sound and ensuring consistency across all audio formats and playback devices (Bogdanov et al., 2019).

Music Recommendation and Discovery

AI's role extends to how we discover music. Streaming platforms like Spotify use AI and machine learning algorithms to analyze users' listening habits and recommend music that aligns with their taste. Spotify's "Discover Weekly" is a popular example of an AI-curated playlist, delivering personalized music recommendations to its users every week (Ricci, Rokach & Shapira, 2015).

Live Performances

Even live music performances are being shaped by AI. Aiva Technologies has created an AI composer that can create orchestral music for live performances. In another fascinating instance, AI was used to bring back the voice of the late rapper Tupac Shakur for a virtual performance at Coachella music festival (Barrat, 2013).

Despite the advances, AI's role in music is complementary to human musicians, rather than replacing them. It serves as a tool that can inspire and enable musicians to explore new creative avenues. The harmony of human creativity with AI technology is what makes the future of music exciting.

Everyday Tech

I. Smartphones

Smartphones are becoming increasingly intelligent due to the infusion of Artificial Intelligence (AI). From personalized recommendations to voice recognition, AI is improving our smartphone experience in numerous ways. It's so ingrained in the smartphone ecosystem that we often don't even realize when we're using it.

Voice Assistants

AI voice assistants such as Apple's Siri, Google's Assistant, and Amazon's Alexa have become commonplace. These AI-powered virtual assistants can perform a variety of tasks like sending messages, making phone calls, setting reminders, and answering questions, all commanded by your voice (Hoy, 2018).

These assistants use a type of AI called natural language processing (NLP), which allows them to understand and respond to human speech. They also employ machine learning to get better at predicting and understanding your needs over time (Hirschberg & Manning, 2015).

Photography

AI has transformed smartphone photography, enhancing the quality of pictures taken with our devices. AI algorithms can identify and enhance specific features in photos, provide optimal lighting, and even blur the background for a professional-looking portrait mode. Google's Pixel phones use AI for their Night Sight feature, which enables clear and detailed low-light photography without the need for flash (Wang et al., 2019).

Personalization

Smartphones use AI to learn from your behaviors and patterns to provide a personalized experience. For instance, predictive text and autocorrect features learn from your typing habits to make smart suggestions, improving the efficiency of your typing (Ruder et al., 2016).

Moreover, app recommendations, news updates, and even your social media feed are all personalized using AI algorithms that learn from your usage patterns and preferences (Ricci, Rokach & Shapira, 2015).

Security

AI also enhances the security of our smartphones. Features like face recognition and fingerprint scanning use AI to provide a secure, personalized unlocking system. AI can also help detect potential threats or malicious activities on your phone, keeping your data safe (Zhou & Jiang, 2012).

AI significantly enhances our smartphone experience, making our devices more user-friendly, personalized, and secure. As AI continues to advance, we can expect our smartphones to become even smarter and more intuitive.

5

"Ethics and responsible AI development should be at the forefront. We need to ensure AI is used for the benefit of all and avoids harm and biases."

- Yoshua Bengio, AI Researcher and Turing Award Winner

Ethical Considerations in AI

Bias in AI: Origins and Impacts

Artificial Intelligence (AI) systems are incredibly powerful, but they're not perfect. One of the most significant challenges facing AI is bias. Bias in AI systems can result in unfair outcomes, ranging from discriminatory hiring practices to injustices in criminal sentencing.

I. What is Bias in AI?

Bias in AI refers to tendencies that AI systems develop which may lead to partial or unfair outcomes. Just like human beings, AI can be biased in its decisions, showing favoritism or prejudice towards particular groups based on factors such as race, gender, or age (Buolamwini & Gebru, 2018).

II. Where Does Bias in AI Come From?

Bias in AI can emerge from several sources:

Data Bias: AI systems, particularly machine learning models, are trained using massive amounts of data. This data often comes from the real world and reflects the biases present in society. For instance, in a dataset used for training a machine learning model to recognize speech, if the vast majority of voices are male, the AI system might become biased towards recognizing male voices more accurately than female ones.

Data bias can also occur when certain groups are overrepresented or underrepresented in the data. For instance, in the field of healthcare, if AI is trained predominantly on data from people of a certain age group, ethnic background, or gender, its ability to accurately diagnose or predict diseases for other demographic groups could be compromised (Chen, Johansson, & Sontag, 2018).

Selection Bias: Selection bias in AI occurs when the dataset used to train an AI model isn't representative of the overall population it's supposed to serve. For example, if a facial recognition system is primarily trained on images of people from one ethnic group, it will be less effective at recognizing individuals from other ethnic groups. This was highlighted in a study by the Massachusetts Institute of Technology, where commercial facial-recognition systems had higher error rates for dark-skinned and female faces (Buolamwini & Gebru, 2018).

Confirmation Bias: Confirmation bias in AI refers to the unconscious incorporation of human prejudices into AI systems. Since humans design and train AI models, it's possible that their conscious or unconscious biases might influence the AI system's behavior. For example, if developers unknowingly favor certain results or outcomes during the AI training process, this could lead to a confirmation bias in the AI system (O'Neil, 2016).

III. Impacts of Bias in AI

Biased AI systems can have significant and detrimental impacts, such as:

Inequality and Discrimination: AI bias can lead to unfair treatment of individuals or groups. For example, a biased hiring algorithm could disadvantage applicants based on their race, gender, or age, leading to discriminatory hiring practices (Dastin, 2018).

Social Consequences: Bias in AI can exacerbate existing social inequalities. For example, if an AI system used in law enforcement disproportionately targets certain demographic groups, it can perpetuate stereotypes and contribute to systemic biases (Richardson, Schultz, & Crawford, 2019).

Loss of Trust: The presence of bias in AI systems can lead to a loss of trust in these systems and the organizations that use them. If people don't trust AI systems due to their biased behavior, they might resist their adoption, hindering technological progress (Rahwan et al., 2019).

Addressing these issues is paramount to ensure the responsible development and deployment of AI. It involves not only technical solutions but also societal awareness, regulation, and active efforts to mitigate bias in data and algorithms. It requires vigilance in data collection, algorithm design, and system deployment, as well as a commitment to transparency and accountability from all those involved in the creation and application of AI systems.

Privacy Concerns:
Data Collection and Use

In the modern digital world, data has become a fundamental resource, sometimes referred to as the 'new oil'. As we've already discussed, AI technologies rely heavily on this 'new oil' to learn, grow, and perform tasks. However, the vast amounts of data collected, often personal data about individuals, raise serious concerns about privacy. Let's explore this issue in more detail.

I. Data Collection

Every time you use an online service, be it a social media platform, an e-commerce site, or a search engine, data is collected about your behavior. This can include what you click on, how long you spend on a page, what you purchase, and even your location. Your interactions with AI technologies, such as voice assistants (like Siri or Alexa) or recommendation algorithms (like those used by Netflix or Amazon), are also recorded, and analyzed.

The primary concern here is the sheer volume and detail of data being collected, often without explicit consent or even the user's knowledge. According to a study by the Pew Research Center, 81% of Americans believe they have little to no control over the data companies collect about them (Auxier et al., 2019).

II. Data Use

Once data is collected, it can be used in a variety of ways, many of which can infringe on privacy. For example, companies might use personal data to target individuals with specific advertisements, a practice known as 'micro-targeting'. While this might simply lead to more relevant ads for some people, others view it as a manipulative practice that takes advantage of their personal information for profit.

Furthermore, collected data can be shared or sold to third parties, further expanding its use. For example, health insurers might be interested in data about a person's lifestyle to adjust insurance rates, a practice that raises serious ethical and privacy concerns (Schwartz & Jahn, 2020).

III. Implications

Privacy concerns related to data collection and use by AI systems can lead to a variety of negative outcomes. These include identity theft, financial fraud, and even potential misuse by governmental bodies (Pasquale, 2015). Moreover, the psychological impact of knowing one's every move could potentially be watched, evaluated, and used for profit can lead to feelings of unease and anxiety, a state sometimes referred to as 'surveillance stress' (Stoycheff, Liu, & Wibowo, 2020).

Addressing privacy concerns in the context of AI is a complex issue. It requires a multi-pronged approach that includes rigorous data protection regulations, robust privacy-preserving technologies, and greater transparency from companies about their data collection and use practices.

Job Displacement: Automation and the Future of Work

AI technologies are revolutionizing our world, bringing efficiency and precision to various tasks across numerous industries. However, along with these improvements come some significant concerns, particularly about the impact of AI and automation on employment. In this section, we'll discuss job displacement, the changing nature of work, and what the future might hold.

I. Automation and Job Displacement

The advancement of AI technology has led to its increasing capability to perform a wide range of tasks, both physical and cognitive, that were traditionally carried out by humans. AI systems are now capable of automating routine physical tasks, such as assembly line work, and cognitive tasks like data analysis and customer service. This trend has raised concerns about the potential impact on jobs and employment.

According to a report by the McKinsey Global Institute, up to 800 million workers globally could be displaced by robotic automation by 2030 (Chui et al., 2017). The report highlights that automation technologies, including AI, have the potential to replace a significant portion of current work activities across various industries.

The capacity of AI systems to learn and improve over time through machine learning algorithms makes them increasingly powerful tools in many sectors. However, this also means that a growing number of jobs may be at risk of being automated. Research conducted by economists Daron Acemoglu and Pascual Restrepo estimates that each additional robot per thousand workers leads to a reduction in the employment-to-population ratio by about 0.18-0.34 percentage points and wages by 0.25-0.5 percent (Acemoglu & Restrepo, 2017).

These findings indicate that the increasing adoption of AI and automation technologies can have significant implications for the workforce. While some jobs may be eliminated or transformed, new job opportunities may also arise as industries adapt to the integration of AI systems. However, the transition can be challenging for individuals whose jobs are at risk of automation, as they may require new skills and training to remain employable in the evolving job market.

It is important for policymakers, businesses, and society as a whole to anticipate and address the potential disruptions caused by AI-driven automation. Strategies such as investing in education and retraining programs, fostering innovation and entrepreneurship, and implementing supportive labor policies can help mitigate the negative impacts and ensure a smooth transition for workers.

Moreover, it is crucial to recognize that AI and automation are not purely job-destroying forces. They have the potential to enhance productivity, improve efficiency, and enable new forms of work and economic growth. By harnessing the power of AI in a responsible and inclusive manner, we can create a future where humans and machines collaborate, leveraging their respective strengths to drive innovation and create meaningful, fulfilling work.

II. Changing Nature of Work

The potential impact of AI on jobs is a topic of significant discussion and concern. While it is true that the widespread adoption of AI technologies may lead to job displacement in certain industries, it is important to view this in the context of historical technological advancements and recognize that AI is likely to change the nature of work rather than completely eliminate it.

Drawing parallels to the Industrial Revolution, which transformed the labor market from predominantly agrarian to factory-based work, the AI Revolution has the potential to bring about a similar shift. New types of jobs, which we cannot fully anticipate today, may emerge as AI technologies advance, and become more integrated into various sectors.

Experts argue that jobs requiring complex decision-making, critical thinking, creativity, and high-level human interaction are less likely to be fully automated by AI. These tasks involve nuanced judgment, emotional intelligence, and the ability to navigate complex social dynamics, which are areas where humans still hold a significant advantage over machines. Instead, AI is more likely to take over routine, repetitive, and mundane aspects of these jobs, allowing humans to focus on more value-added tasks (Bessen, 2019).

For example, in the healthcare sector, AI can assist doctors by analyzing medical images, diagnosing diseases, and suggesting treatment options. While AI can provide valuable insights and support, it is ultimately up to the human healthcare professionals to make critical decisions based on their expertise and patient interaction.

Similarly, in the creative industry, AI tools can aid in generating ideas, automating repetitive design tasks, or even composing music. However, the ultimate creative vision and artistic interpretation still rely on human ingenuity and innovation.

It is worth noting that as AI technology progresses, the nature of work will continue to evolve, and individuals will need to adapt and acquire new skills to remain relevant in the job market. This calls for a focus on lifelong learning, reskilling, and upskilling to ensure that individuals are equipped with the necessary capabilities to thrive in an AI-driven world.

While concerns about job displacement are valid, history has shown that technological advancements have often resulted in the transformation and creation of new types of jobs. AI is likely to change the nature of work, with humans and machines collaborating in complementary ways. By leveraging the capabilities of AI to automate routine tasks, humans can engage in more meaningful, complex, and high-value work, ultimately shaping a future where AI enhances human potential rather than replacing it.

III. The Future of Work

The future of work in the era of AI is a topic of significant concern and debate. On one hand, there is the potential for widespread job displacement as AI systems increasingly take over routine and repetitive tasks. This could lead to social and economic upheaval, as workers in affected industries face unemployment or the need to transition into new roles. On the other hand, the emergence of AI also holds the promise of creating new job opportunities and driving economic growth.

To navigate this complex landscape, proactive steps need to be taken by governments, industries, and educational institutions. First and foremost, investing in retraining and upskilling programs is crucial to equip workers with the skills needed for the jobs of the future. By focusing on areas that are less likely to be automated, such as complex decision making, critical thinking, creativity, and high-level human interaction, individuals can position themselves for the changing nature of work (Bessen, 2019).

Additionally, social safety nets must be put in place to support workers who are displaced by automation. This may involve providing financial assistance, access to healthcare, and other forms of support to help individuals transition into new careers or industries. A robust safety net can help alleviate the social and economic challenges associated with job displacement and ensure a more equitable transition (Furman et al., 2019).

Furthermore, creating a regulatory environment that encourages innovation while addressing the potential negative impacts of AI is essential. This includes considering issues such as worker rights, privacy protection, and algorithmic transparency. Striking a balance between fostering innovation and safeguarding the well-being of workers and society as a whole is key to a successful and sustainable AI-driven future (Brynjolfsson & McAfee, 2017).

As economist Carl Frey aptly notes, historical transitions, such as the Industrial Revolution, brought both significant benefits and adjustment costs. Similarly, the future of work with AI will likely entail challenges and opportunities. Frey emphasizes the need to prepare for a future in which machines increasingly perform routine tasks, highlighting the importance of proactive measures and planning for the inevitable changes ahead (Frey, 2019).

The future of work in the age of AI is a complex and multifaceted topic. It presents challenges related to job displacement, social inequality, and economic disruption, but also offers opportunities for innovation, economic growth, and the creation of new types of jobs. By investing in retraining, establishing social safety nets, and fostering a regulatory environment that balances innovation and worker protection, we can navigate this transformative period and strive for a future where humans and machines coexist and thrive.

AI and Warfare: Autonomous Weapons and Ethical Dilemmas

As AI technology advances, it's being integrated into many sectors, including military and defense. However, the use of AI in warfare, particularly in the development of autonomous weapons, raises complex ethical dilemmas. In this section, we'll delve into these concerns and their potential implications.

I. Autonomous Weapons

Autonomous weapons, often known as 'killer robots,' are a controversial development in the field of AI and military technology. These weapons systems are designed to operate without direct human control, using AI algorithms to identify, track, and engage targets with high precision and speed. While proponents argue that autonomous weapons can enhance military capabilities and reduce risks to human soldiers, critics raise concerns about the ethical implications and potential for unintended consequences.

One notable example of autonomous weapon systems is the US Navy's Aegis Combat System. This advanced naval defense system integrates AI technology to autonomously detect, track, and intercept incoming missiles and aircraft. The system utilizes radar, sensors, and AI algorithms to rapidly analyze data and make split-second decisions on engaging targets. The Aegis Combat System demonstrates the potential of AI-powered autonomous weapons to enhance military capabilities by providing advanced defense capabilities with reduced response times (United States Navy, 2021).

Another example is the SGR-A1 sentry gun deployed by South Korea. This AI-powered system is designed to perform surveillance and defense functions along the Korean Demilitarized Zone. Equipped with various sensors, including thermal imaging and voice recognition, the SGR-A1 can detect and track potential threats autonomously. The system has the capability to make decisions on target engagement and can open fire without human intervention if deemed necessary. The deployment of the SGR-A1 highlights the growing use of autonomous weapons for border security and surveillance purposes (Kania, 2021).

The development and deployment of autonomous weapons raises significant ethical concerns and policy debates. Critics argue that removing human control from the decision-making process can lead to unintended consequences, such as mistaken targeting or disproportionate use of force. There are concerns about the potential for these weapons to violate principles of humanitarian law, including the distinction between combatants and non-combatants, and the principle of proportionality. The lack of human judgment and accountability in the use of force also raises issues of responsibility and liability.

International organizations, including the United Nations, have been discussing the ethical and legal implications of autonomous weapons. Efforts are underway to establish frameworks and guidelines for the responsible use of these technologies. The Campaign to Stop Killer Robots, a coalition of NGOs, is advocating for a global ban on fully autonomous weapons to prevent the potential risks associated with their deployment.

The development of autonomous weapons is a complex and multifaceted issue. It requires careful consideration of legal, ethical, and humanitarian concerns to ensure that the use of AI in military applications aligns with international norms and principles. Striking a balance between leveraging technological advancements for defense purposes and maintaining human control and ethical decision-making in the use of force is essential to address the challenges posed by autonomous weapons.

II. Ethical Dilemmas

The development and deployment of autonomous weapons raises profound ethical questions that demand careful consideration. One of the primary concerns is the issue of accountability. When an autonomous weapon causes unintended harm or civilian casualties, determining who should be held responsible becomes challenging. Should it be the creators of the AI technology, the military commanders who deployed the system, or the autonomous weapon itself? This question raises complex legal and moral dilemmas that require thoughtful examination and clear frameworks for assigning responsibility (Russell et al., 2015).

Another significant concern is the potential for an "AI arms race" among nations. As countries compete to develop increasingly advanced autonomous weapons, there is a risk of escalating conflicts and the emergence of a new kind of warfare that is primarily conducted by autonomous systems. In such a scenario, decision-making processes guided by human compassion, empathy, and moral judgment may be absent, raising concerns about the ethical implications of warfare devoid of human involvement (Russell et al., 2015).

The third concern relates to the risk of autonomous weapons falling into the wrong hands. The relative ease of producing software-based technologies compared to traditional weapons systems increases the potential for terrorists and rogue states to acquire and misuse AI-powered weapons. This presents a significant challenge in terms of international security and the potential for these weapons to cause widespread harm without the necessary safeguards and controls in place (Horowitz & Scharre, 2015).

Addressing these ethical concerns requires a comprehensive and multidisciplinary approach. International dialogue and cooperation are essential to establish frameworks and norms that govern the development, deployment, and use of autonomous weapons. The involvement of experts in AI, robotics, law, ethics, and international relations is crucial to navigate the complex challenges and ensure that autonomous weapons are developed and utilized in a manner that aligns with ethical principles, international law, and human rights standards.

III. Call for Regulation

The ethical concerns surrounding autonomous weapons have prompted a strong call for international regulation from various stakeholders. AI researchers, human rights activists, and industry leaders have joined forces to advocate for strict controls on the development and use of these weapons. The Future of Life Institute, in particular, played a significant role in raising awareness and mobilizing support for regulations.

In an open letter published by the Future of Life Institute, thousands of signatories, including renowned physicist Stephen Hawking and SpaceX CEO Elon Musk, called for a ban on offensive autonomous weapons that operate beyond meaningful human control. This appeal emphasizes the need to ensure that human judgment and accountability remain central to the decision-making process in warfare (Future of Life Institute, 2015).

The push for regulation recognizes the potential benefits that AI can bring to military operations, such as enhanced capabilities and increased protection for soldiers. However, it also underscores the crucial importance of addressing the profound ethical challenges associated with the use of AI in warfare.

Effectively navigating these challenges requires a multifaceted approach. It necessitates thoughtful dialogue and engagement among stakeholders, including AI researchers, policymakers, legal experts, military personnel, and human rights advocates. These discussions should focus on establishing rigorous ethical guidelines that govern the development, deployment, and use of autonomous weapons.

International cooperation is key to achieving meaningful regulation. Collaboration among nations, international organizations, and non-governmental entities is essential to establish frameworks that ensure the responsible and ethical use of AI technologies in military contexts. This cooperation can facilitate the exchange of best practices, the harmonization of standards, and the enforcement of regulations that prioritize human safety, minimize civilian harm, and maintain compliance with international humanitarian law.

Ultimately, the goal is to strike a balance between leveraging the potential benefits of AI in military operations while upholding ethical principles and safeguarding human rights. By promoting open dialogue, rigorous ethical guidelines, and international cooperation, we can strive to navigate the complex landscape of autonomous weapons and shape a future where AI technologies are deployed in a manner that aligns with human values and global security.

6

"The potential of AI is limited only by our imagination. It will redefine industries, create new opportunities, and shape the future of humanity."

- Demis Hassabis, CEO of DeepMind

The Future of AI: Possibilities and Challenges

Potential Advancements

I. Artificial General Intelligence (AGI)

As we move further into the world of AI, a significant development on the horizon is Artificial General Intelligence (AGI). Unlike the narrow AI systems, we have today, AGI refers to a type of AI that possesses the ability to understand, learn, and apply knowledge across a wide range of tasks at a level equal to or even surpassing that of a human.

To recap, Artificial General Intelligence, also known as "strong AI" or "full AI", is essentially a machine with the ability to perform any intellectual task that a human being can do. It would be able to reason, solve puzzles, make judgments under uncertainty, plan, learn from experience, and adapt to new situations (Goertzel & Pennachin, 2007).

AGI would possess not just narrow expertise in specific tasks, but also a broad understanding that can be applied across domains. It would have the ability to transfer learning from one domain to another, a feat which current AI struggles with. In essence, an AGI system would be capable of understanding or learning any intellectual task that a human being can.

Future Prospects and Practical Uses

Imagine a huge leap forward in what our current AI systems can do. That's what we could expect from the emergence of Artificial General Intelligence (AGI). This is a type of AI that could not only match human performance but potentially surpass it in most tasks that have value in our economy.

Think of it like this: instead of an AI that excels only at specific tasks, we'd have an AI whizz that can turn its 'hand' to pretty much anything, just like you or me. It could work faster, without taking breaks, and complete tasks more efficiently than a human ever could.

Take scientific research as an example. It involves pouring over huge volumes of data, a time-consuming task for any human. An AGI system could sift through this information at lightning speed, spotting new patterns and making discoveries that might take humans years, or that we might miss entirely.

Here's another way to look at it. Right now, narrow AI (the type of AI we currently have) can diagnose diseases based on a set of symptoms, a bit like a really clever interactive medical textbook. But AGI could take this a step further. Imagine it as a world-class doctor with an encyclopedic knowledge of every medical development, and access to a vast database of patient histories. It could assess your symptoms, consider your medical history, consider the latest research, and even consider socio-economic factors that might impact your health. It could then use all this information to devise a treatment plan tailored precisely for you (Bostrom, 2014).

In a nutshell, the advent of AGI could revolutionize fields like medicine, science, and beyond, taking our current use of AI to entirely new levels.

Challenges and Ethical Concerns

However, AGI also comes with significant challenges and ethical concerns. The development of AGI would have profound implications for society, and it's important that we navigate this path with caution.

One key concern is the problem of control: if an AGI system becomes more intelligent than humans, would we be able to control it? And how can we ensure that it aligns with human values and goals? These questions are the focus of a field of research called AI alignment (Russell, Dewey, & Tegmark, 2015).

Furthermore, there are concerns about AGI leading to extreme economic inequality, as those with access to AGI could potentially gain a disproportionate amount of wealth and power. There's also the risk of job displacement on a large scale, as AGI could outperform humans at most tasks (Brynjolfsson & McAfee, 2014).

While AGI represents a significant potential advancement in AI, it also brings substantial challenges and ethical concerns. As we continue to explore the path to AGI, it's crucial that we do so responsibly, considering both the potential benefits and the risks involved.

Exciting Prospects: Quantum Computing

Quantum computing sounds like something out of a sci-fi movie, right? But it's a real field of study that could radically change our world, particularly in the realm of AI.

Before we get into the 'how', let's start with the 'what'. In simple terms, quantum computers are machines that use the principles of quantum physics to process information. Quantum physics, also known as quantum mechanics, is the part of physics that deals with the tiniest particles in the universe, like atoms and photons.

Now, if you're thinking, "Wait, don't all computers process information?" You're right. But quantum computers do it in a way that's fundamentally different from classical computers. Classical computers, like the laptop or smartphone you're probably reading this on, process information in bits, which can be either a 0 or a 1. This is like flipping a coin; it can land either heads up or tails up.

Quantum computers, on the other hand, process information in quantum bits, or qubits. A qubit can be both 0 and 1 at the same time, thanks to a quantum physics principle called superposition. If you're thinking this sounds like flipping a coin and having it land both heads up and tails up at the same time, you're getting the idea.

So why does this matter? Well, thanks to superposition, and another quantum principle called entanglement, quantum computers can process vast amounts of data simultaneously. This makes them potentially exponentially more powerful than classical computers, opening up possibilities we can't even fully comprehend yet.

As it relates to AI, the increased processing power of quantum computers could enable us to create far more complex and sophisticated AI models than we can with classical computers. This could lead to breakthroughs in fields like machine learning, natural language processing, and data analysis, significantly advancing our AI capabilities (Biamonte et al., 2017).

However, it's important to note that quantum computing is still in its infancy. There are significant technical challenges to overcome before we have practical, large-scale quantum computers. But the potential is there, and it's a fascinating area to watch.

The Next Frontier: AI in Space Exploration

When we think of space exploration, we often picture astronauts bravely journeying into the unknown. However, the future of space exploration may look a little different, largely thanks to advancements in AI.

AI can be a game-changer for space exploration in many ways, making missions safer, more efficient, and more likely to succeed. One of the key benefits of AI is its ability to process and analyze massive amounts of data quickly and accurately. In the context of space exploration, this could be applied to tasks like analyzing images of planetary surfaces to identify features of interest or monitoring the health and performance of spacecraft systems to detect and resolve issues before they become critical (Fong, Bualat, Deans & Haggy, 2019).

AI could also be used to autonomously navigate spacecraft or rovers, allowing them to avoid hazards and reach their destinations more efficiently. For instance, NASA's Perseverance

Rover, which landed on Mars in February 2021, is equipped with an AI-powered navigation system that allows it to avoid obstacles and navigate its environment without constant input from mission control (NASA, 2021).

Even more exciting are the potential applications of AI in the search for extraterrestrial life. By analyzing the data collected from telescopes and spacecraft, AI algorithms could help identify planets that may be suitable for life, or even detect signs of life itself. For instance, researchers at the University of Texas at Austin used machine learning to analyze data from the Kepler Space Telescope, discovering two exoplanets that had previously been missed by human analysts (Shallue & Vanderburg, 2018).

Despite these exciting prospects, it's important to note that there are still significant challenges to overcome. Extreme conditions in space, including radiation and lack of power, can be damaging to electronic systems, including those used by AI. Additionally, communication delays between Earth and spacecraft make it difficult to control and update AI systems in real time, requiring them to operate with a high degree of autonomy (Fong et al., 2019).

Nevertheless, the potential for AI to revolutionize space exploration is clear, and it's an area of research that is likely to continue to grow and evolve in the coming years.

Ten Advancements Using AI In Space Exploration

I. Autonomous Navigation

AI can help spacecraft and rovers navigate unfamiliar terrains autonomously. It can quickly process information about the environment and make split-second decisions to avoid hazards, which is particularly beneficial in distant locations like Mars where communication delays can be significant. NASA's Perseverance Rover uses an AI-based navigation system for this purpose (NASA, 2021).

II. Planet Habitability Analysis

AI can analyze data from telescopes to identify planets outside our solar system (exoplanets) that may be habitable. Machine learning algorithms can comb through vast amounts of data to spot potential signs of habitable conditions, such as a suitable atmosphere and presence of water (Shallue & Vanderburg, 2018).

III. Discovery of New Astronomical Bodies

AI can aid in the discovery of new astronomical bodies, such as asteroids, comets, or even unknown planets. It can sift through large datasets from space observatories to detect patterns or anomalies that humans might miss (Pearson et al., 2019).

IV. Spacecraft Health Monitoring

AI can monitor the health and performance of spacecraft systems. It can predict potential faults or failures by analyzing data on the functioning of various components, thereby preventing critical system failures (Chien et al., 2018).

V. Satellite Image Analysis

AI can be used to analyze satellite images of Earth for various purposes, such as tracking climate change, monitoring natural disasters, or mapping land use. AI can process these images faster and more accurately than humans, making it a valuable tool for Earth observation (Jean et al., 2016).

VI. Astrobiology Research

AI can aid in the search for extraterrestrial life by analyzing data from various space missions and telescopes. Machine learning algorithms can identify promising areas or signals for further investigation (Cabrol, 2018).

VII. Deep Space Communication

AI could potentially optimize deep space communication, ensuring that signals between distant spacecraft and Earth are received and decoded correctly, despite the vast distances and interference from cosmic phenomena (Noffz et al., 2019).

VIII. Space Traffic Management

With the increasing number of satellites and space debris in Earth's orbit, AI could be used to track and predict the paths of these objects to avoid collisions. This is crucial for ensuring the safety and longevity of satellites and space stations (Fujimoto et al., 2020).

IX. Robotic Assembly and Repairs in Space

AI could enable robots to carry out assembly, maintenance, and repairs of spacecraft or space structures, reducing the need for risky spacewalks by astronauts. NASA's Robonaut 2 has been designed with this idea in mind (Diftler et al., 2011).

X. Advanced Scientific Research

AI could help analyze vast amounts of data gathered from space, leading to new insights about the universe. This could include everything from understanding star formation to the structure of galaxies (Baron & Poznanski, 2017).

Potential Risks

I. Superintelligent AI

In simple terms, superintelligent AI is a form of artificial intelligence that surpasses human intelligence in practically every field, including scientific creativity, general wisdom, and social skills. Today, our AI technologies, although advanced, are not near this superintelligent level. They are specialized, meaning they excel in tasks they are designed for but are clueless outside of them. However, the possibility of developing a superintelligent AI is not ruled out by researchers, raising concerns about the risks it could pose (Bostrom, 2014).

Control Problem

The first and foremost risk is the control problem, as mentioned by philosopher and researcher Nick Bostrom. It's the idea that if we create an AI significantly more intelligent than us, we might not be able to control it. It could develop its own goals, and if they don't align with ours, it could lead to undesirable outcomes (Bostrom, 2014).

Existential Risk

The control problem leads to what is known as the existential risk. If a superintelligent AI's goals conflict with our survival and well-being, it could cause human extinction or any outcome that would permanently and drastically curtail humanity's potential. Elon Musk famously stated, "We are summoning the demon" in reference to our pursuit of superintelligent AI (Musk, 2014).

Economic Disparity and Job Displacement

Another risk is the exacerbation of economic disparity. Superintelligent AI could lead to significant job displacement and wealth concentration among those who own and control these AI technologies. In the wrong hands, superintelligent AI could be used in ways that harm humanity or unduly concentrate power (Brynjolfsson & McAfee, 2014).

AI Arms Race

The potential for an AI arms race poses another risk. In the rush to achieve superintelligence, safety precautions could be overlooked, leading to the development of AI that isn't aligned with human values (Russell et al., 2015).

Autonomy and Accountability

Finally, there's a risk tied to the autonomy of superintelligent AI. If an autonomous AI makes decisions resulting in harm, determining accountability can be challenging. This issue becomes more complex with superintelligent AI that can change and rewrite its code (Roff & Moyes, 2016).

While these risks might seem to belong to the distant future, it is crucial to consider them now. Thoughtful regulation and research dedicated to aligning AI's goals with ours are crucial. As Stephen Hawking noted, "The rise of powerful AI will be either the best or the worst thing ever to happen to humanity. We do not yet know which" (Cellan-Jones, 2014).

II. Dependence on AI

As our society increasingly integrates AI into our daily lives, from autonomous vehicles to virtual assistants, we're growing more dependent on these technologies. While AI has brought numerous benefits, this dependence isn't without risks, which we will explore in the following subsections.

Reliability and Trust

Dependency on AI systems comes with inherent risks. As we increasingly rely on AI to make critical decisions and automate various tasks, we place a significant amount of trust in their accuracy and performance. However, AI systems are not infallible, and they can make mistakes that can have serious consequences.

One of the concerns with AI dependency is in the realm of autonomous vehicles. Tesla's autopilot system, for example, has faced scrutiny and criticism due to incidents where the system failed to detect and respond appropriately to certain scenarios. These incidents have tragically resulted in fatal accidents. It highlights the importance of understanding the limitations and capabilities of AI systems, as well as the responsibility of users to be aware of their roles in operating such systems.

Another area where dependency on AI can pose risks is in critical decision-making processes, such as in healthcare or finance. AI algorithms used in medical diagnosis or financial predictions, for instance, may generate incorrect results or recommendations that could have severe implications for patient care or financial stability.

Moreover, biases and limitations in AI algorithms can also contribute to the risks associated with AI dependency. If the training data used to develop AI models is biased or lacks diversity, it can result in biased outcomes, perpetuating societal inequalities or reinforcing existing biases.

To address these risks, it is essential to establish robust testing and validation processes for AI systems to ensure their reliability and safety. Ongoing monitoring and evaluation of AI systems' performance, as well as transparent reporting of errors and limitations, can help identify and rectify potential issues. Additionally, promoting user education and awareness about the capabilities and limitations of AI systems can prevent misunderstandings and inappropriate reliance on their functionalities.

As AI continues to evolve and integrate into various aspects of our lives, it is crucial to strike a balance between the benefits and risks. Responsible development, regulation, and continuous improvement of AI systems are vital to mitigate potential risks and ensure their safe and ethical use in society.

Security Vulnerabilities

The dependency on AI systems also brings about new security vulnerabilities that need to be addressed. As AI becomes more prevalent in various domains, cybercriminals are finding ways to exploit these systems through adversarial attacks. Adversarial attacks involve manipulating the input data fed into AI algorithms to deceive the system and cause it to make incorrect decisions.

One specific type of adversarial attack is known as a "poisoning attack." In this attack, an attacker injects malicious or misleading data into the training dataset used to develop AI models. By introducing subtle changes to the input data, such as adding noise or modifying pixels in images, the attacker can manipulate the AI system's behavior.

For example, in the context of facial recognition, adversarial attacks can be used to trick the system into misclassifying a person's identity or granting unauthorized access. By altering or disguising certain facial features, an attacker can deceive the AI system and bypass security measures.

Similarly, adversarial attacks on autonomous driving systems pose serious risks. By subtly modifying traffic signs or road markings, an attacker can fool the AI system into misinterpreting the environment, potentially leading to accidents or unsafe driving decisions.

These adversarial attacks highlight the importance of ensuring the robustness and security of AI systems. Researchers and developers are actively working on developing defenses against such attacks, including improving the resilience of AI algorithms and training models to detect and mitigate adversarial manipulations.

Furthermore, ongoing research and collaboration between AI practitioners, cybersecurity experts, and policymakers are necessary to establish guidelines and standards for securing AI systems. Regular testing and auditing of AI models for vulnerabilities, as well as implementing measures to detect and prevent adversarial attacks, are crucial steps to protect against these risks.

As AI technology continues to advance, it is essential to stay vigilant and proactive in addressing the security risks associated with AI dependency. By implementing robust security measures and fostering a culture of cybersecurity awareness, we can mitigate the potential harms and ensure the safe and responsible use of AI systems.

Loss of Skills

The increased dependency on AI systems carries the risk of a potential loss of essential skills among individuals. As AI takes over various tasks and responsibilities, there is a possibility that humans may become less proficient or even lose the skills that were once crucial for performing those tasks manually.

In aviation, for instance, pilots have traditionally been trained to manually fly aircraft and handle various flight operations. However, with the advancement of automated flight systems, pilots now heavily rely on autopilot and other AI-based systems to handle many aspects of flight control. This shift towards automation has raised concerns about the potential degradation of pilots' manual flying skills.

Studies and reports from aviation authorities, such as the European Union Aviation Safety Agency (EASA), have highlighted the need for pilots to maintain and regularly practice their manual flying skills to ensure they can handle unexpected situations or system failures effectively. Over-reliance on automated systems can lead to a loss of proficiency in manual flying techniques, reducing pilots' ability to take control when needed.

The risk of skill degradation is not limited to aviation alone. In various domains where AI systems are deployed, individuals may experience a decline in skills associated with tasks that are now performed by AI. For example, as automated translation systems become more advanced, individuals may rely less on their own language proficiency and translation abilities, leading to a potential erosion of language skills.

To mitigate this risk, it is crucial to strike a balance between the use of AI systems and the preservation of essential human skills. It involves designing AI systems that encourage human involvement and skill development rather than completely replacing human capabilities. Additionally, continuous training and education programs can help individuals adapt to AI technologies and ensure they maintain essential skills that are still valuable in situations where AI may not be fully reliable or available.

Ultimately, a thoughtful approach is needed to leverage AI systems while preserving and nurturing the unique skills and capabilities that humans bring. It involves recognizing the importance of maintaining a diverse skill set, promoting ongoing learning, and ensuring that AI is used as a complementary tool rather than a complete substitute for human expertise.

Economic Displacement

The rapid advancement of AI technologies poses a significant risk in terms of job dependency. While traditionally it was routine and repetitive tasks that were most susceptible to automation, the current trajectory of AI development suggests that a wide range of jobs across various industries may be at risk. This includes not only manual labor but also cognitive and non-routine tasks that were previously considered safe from automation.

The increasing reliance on AI for economic productivity can potentially lead to significant job displacement and, in turn, create social unrest. As AI systems become more sophisticated and capable of performing complex tasks, the need for human labor in certain areas may diminish. This could result in widespread unemployment or underemployment, with individuals struggling to find suitable alternative employment opportunities.

Research conducted by Arntz, Gregory, and Zierahn (2016) indicates that a substantial number of jobs across different skill levels are vulnerable to automation in the coming decades. Their study suggests that up to 47% of total US employment is at risk of automation. This figure includes not only low-skilled jobs but also occupations in the middle and high-skill ranges, such as administrative tasks, data analysis, and even some aspects of professional work.

The potential consequences of widespread job displacement are significant. Apart from the economic impact on individuals and families, the loss of jobs can lead to social unrest and a widening wealth gap. It may exacerbate inequality and create challenges in ensuring social stability and well-being.

Addressing the risks associated with job dependency on AI requires proactive measures. This includes preparing individuals for the changing job landscape through education and training programs that focus on developing skills that are less susceptible to automation. Governments, educational institutions, and industries need to collaborate to identify emerging job opportunities and provide reskilling and upskilling initiatives to help individuals transition into new roles.

Furthermore, fostering an environment that supports entrepreneurship and innovation can help create new job opportunities that align with the evolving needs of the AI-driven economy. Policies that promote job creation and support the development of industries that leverage AI technologies responsibly and ethically can mitigate the potential negative impacts of job displacement.

The growing dependence on AI technologies for economic productivity carries the risk of significant job displacement and potential social unrest. To mitigate these risks, it is crucial to invest in education and training, promote entrepreneurship, and implement policies that facilitate a smooth transition into the changing job landscape. By taking proactive steps, societies can better prepare for the impact of AI on employment and ensure a more equitable and prosperous future.

Ethical and Privacy Concerns

Lastly, reliance on AI systems introduces significant ethical and privacy concerns that arise from the collection, storage, and use of vast amounts of data. AI systems depend on data to learn and make predictions, and this reliance can create risks regarding the ethical use and protection of that data.

One of the main concerns is the potential for misuse or unauthorized access to sensitive information. AI systems often require access to personal data, such as user profiles, browsing history, or health records, to provide personalized services or make accurate predictions. However, the collection and utilization of this data can be subject to abuse or breaches, leading to privacy violations and potential harm to individuals.

Moreover, the scale and complexity of AI systems make it challenging to ensure that data is used ethically and responsibly. AI algorithms learn patterns and make decisions based on the data they are trained on, but if the data contains biases, discrimination, or inaccuracies, these issues can be perpetuated or amplified by the AI system. For example, if a facial recognition system is trained on biased or unrepresentative datasets, it may result in discriminatory outcomes, such as misidentifying individuals from certain ethnic backgrounds.

The work of Zuboff (2019) highlights the concerns around data privacy and the potential for surveillance capitalism, where AI systems are used to exploit personal data for commercial gain. This raises questions about individual autonomy, consent, and the control individuals have over their personal information in an AI-driven society.

To mitigate these risks, it is essential to establish robust ethical frameworks and regulations to govern the collection, use, and protection of data in AI systems. Transparency and accountability mechanisms should be implemented to ensure that individuals understand how their data is being used and have control over its usage. Additionally, privacy-preserving techniques, such as data anonymization and encryption, can help safeguard sensitive information while still allowing AI systems to derive meaningful insights.

Furthermore, promoting ethical practices and responsible AI development within organizations and industries is crucial. This includes adhering to principles of fairness, transparency, and accountability throughout the AI lifecycle, from data collection to algorithm design and deployment. Regular audits and independent assessments can help ensure compliance with ethical standards and identify potential risks or biases in AI systems.

Reliance on AI systems brings forth ethical and privacy concerns due to the collection and utilization of large amounts of data. Safeguarding personal information, addressing biases, and ensuring transparency and accountability are vital to mitigate these risks. Ethical frameworks, privacy regulations, and responsible practices must be established to ensure that AI systems are developed and deployed in a manner that respects individuals' rights, privacy, and autonomy.

While AI holds incredible potential to enhance our lives, our growing dependence on it is not without risks. As we continue to integrate AI into our society, it's crucial to address these challenges, ensuring AI develops in a way that is safe, ethical, and beneficial for all.

III. The Cybersecurity Landscape in the AI Era

Artificial Intelligence (AI) brings a new dimension to cybersecurity. While AI can be an invaluable tool in enhancing our defenses, it also introduces new threats and vulnerabilities. The increasing dependence on AI systems across a wide range of sectors makes cybersecurity a pressing issue.

AI as a Cybersecurity Tool

AI indeed holds tremendous potential as a cybersecurity tool and has revolutionized the way we approach threat detection and response. By leveraging advanced algorithms and machine learning techniques, AI-powered cybersecurity systems can analyze vast quantities of data, detect anomalies, and identify potential threats more quickly and accurately than human analysts alone.

One of the primary advantages of AI in cybersecurity is its ability to process and analyze large volumes of data in real-time. Traditional methods of cybersecurity often involve manual analysis of logs and data, which can be time-consuming and prone to human error. AI, on the other hand, can swiftly sift through massive datasets, identify patterns, and detect anomalies that might indicate a cyberattack or a security breach.

AI-powered cybersecurity systems use machine learning algorithms to train large datasets of known threats and normal behavior patterns. This enables them to recognize and understand common attack vectors, such as malware signatures, network anomalies, or unusual user behavior. By continuously learning and adapting to new threats, AI systems can evolve and improve their detection capabilities over time.

For instance, AI algorithms can monitor network traffic and detect suspicious patterns that may indicate a potential cyberattack, such as an unusually high volume of data transfers or unauthorized access attempts. They can also identify known attack signatures and malware patterns, allowing for immediate detection and response.

Moreover, AI can help in automating response actions, enabling a swift and proactive defense against cyber threats. AI-powered systems can automatically block or quarantine suspicious activities, initiate incident response procedures, and provide real-time alerts to cybersecurity teams, enabling them to take immediate action to mitigate the potential impact of an attack.

AI's potential as a cybersecurity tool is undeniable. Its ability to analyze vast amounts of data, detect anomalies, and respond swiftly offers a significant advantage in the ever-evolving threat landscape. By leveraging AI-powered cybersecurity systems, organizations can strengthen their defenses, detect attacks in real-time, and respond effectively to mitigate potential damage. However, a balanced approach that combines AI with human expertise is necessary to maximize the effectiveness and reliability of cybersecurity operations.

AI-Driven Cyberattacks

Conversely, while AI has immense potential as a cybersecurity tool, it also presents significant risks when in the hands of cybercriminals. The same capabilities that make AI a potent defense mechanism can be exploited to launch sophisticated and targeted cyberattacks.

One concerning aspect is the use of AI in developing advanced attack techniques. Cybercriminals can harness AI algorithms to automate and optimize their attack strategies. For example, AI-powered systems can be trained to scan vast amounts of data to identify vulnerabilities in software or network systems more efficiently than human attackers. This allows them to exploit weaknesses and launch targeted attacks with greater precision and effectiveness.

Furthermore, AI can be employed to develop convincing social engineering techniques. With the rise of deepfake technologies, AI algorithms can create realistic audio or video impersonations of trusted individuals. This could be used to manipulate or deceive individuals into providing sensitive information, such as passwords or financial data. By mimicking the voice or appearance of someone familiar, cybercriminals can increase their chances of successfully infiltrating systems or tricking individuals into divulging confidential information.

Additionally, AI can be utilized to automate and optimize the process of crafting phishing emails or malware. AI algorithms can generate and personalize convincing messages, making them more difficult for traditional email filters to detect. This increases the chances of unsuspecting individuals falling victim to phishing attacks or inadvertently downloading malware.

Another concern is the potential use of AI to evade detection by security systems. Adversarial attacks, where subtle modifications are made to input data to deceive AI algorithms, can be employed to bypass security measures. By exploiting vulnerabilities in AI models, attackers can evade detection and execute malicious activities without triggering alarms or raising suspicion.

To counter these evolving threats, cybersecurity professionals must develop robust defenses and constantly update their systems to adapt to new attack techniques. This includes employing AI-powered security solutions that can detect and mitigate AI-driven attacks. By leveraging AI against AI, security systems can enhance their ability to identify and respond to malicious activities.

Furthermore, collaboration between industry, academia, and government is crucial to stay ahead of cybercriminals. Sharing knowledge, research, and best practices can help develop effective countermeasures against AI-driven attacks. Additionally, ethical frameworks and regulations should be established to guide the responsible use of AI technologies and prevent their misuse in malicious activities.

The same capabilities that make AI a powerful defense mechanism can be exploited by cybercriminals to launch sophisticated attacks. To address this, a proactive approach that includes advanced security measures, collaboration, and responsible use of AI is essential to stay ahead of evolving threats and protect against AI-driven cybercrime.

Vulnerability of AI Systems

AI systems, despite their remarkable capabilities, are not immune to cyberattacks. Malicious actors can specifically target AI systems using techniques known as adversarial attacks. These attacks aim to manipulate the input data fed to AI models, leading to incorrect outputs or system malfunctions.

Adversarial attacks exploit the vulnerabilities and limitations of AI algorithms by introducing carefully crafted perturbations or modifications to the input data. These perturbations are often imperceptible to humans but can have a significant impact on the behavior and performance of AI models.

One type of adversarial attack is called the evasion attack. In this scenario, an attacker intentionally modifies the input data to mislead the AI system into making incorrect decisions or classifications. For example, an attacker could manipulate an image by adding imperceptible perturbations that cause an AI-based image recognition system to misclassify the image.

Another type of adversarial attack is the poisoning attack. In this case, an attacker manipulates the training data used to train an AI model. By injecting malicious or misleading examples into the training dataset, the attacker aims to compromise the model's performance or introduce vulnerabilities that can be exploited later. For instance, an attacker could add manipulated training samples to a dataset used for training a spam email filter, tricking the AI system into classifying legitimate emails as spam.

Adversarial attacks pose a significant challenge in ensuring the robustness and reliability of AI systems. These attacks can have severe consequences, such as causing autonomous vehicles to misinterpret road signs or manipulating the decision-making process of AI systems used in critical applications like healthcare or finance.

Addressing adversarial attacks requires ongoing research and the development of defense mechanisms. One approach is to enhance the resilience of AI models by employing techniques such as adversarial training, where models are trained using both clean and adversarial examples to make them more robust against manipulation. Additionally, techniques like input sanitization and anomaly detection can be used to identify and mitigate adversarial attacks.

To ensure the security of AI systems, it is crucial to implement robust testing and validation procedures. Rigorous evaluation and stress testing can help identify vulnerabilities and potential weaknesses in AI models, allowing for the implementation of appropriate safeguards.

Furthermore, ongoing collaboration between researchers, industry professionals, and policymakers is essential to stay ahead of evolving adversarial attacks. Sharing knowledge, developing standardized evaluation methodologies, and establishing best practices can help mitigate the risks posed by adversarial attacks on AI systems.

Privacy Implications

The reliance of AI systems on large datasets introduces significant privacy concerns. In order to train and operate effectively, AI algorithms often require access to vast amounts of data, including personal and sensitive information. This data can include personal records, financial transactions, browsing history, social media posts, and more.

The collection and use of such data raise concerns about how it is handled, stored, and protected. If AI systems are not properly secured, they can become targets for cyberattacks, potentially leading to data breaches and the exposure of sensitive information. This can have severe consequences for individuals, including identity theft, financial fraud, reputational damage, and invasion of personal privacy.

One example of the privacy risks associated with AI is the increasing use of facial recognition technology. Facial recognition algorithms analyze and process facial images to identify individuals, but this raises concerns about the potential misuse of biometric data. If these systems are compromised or fall into the wrong hands, sensitive facial data could be exploited for unauthorized surveillance or tracking purposes.

Moreover, AI systems that rely on cloud computing or data sharing among different entities further amplify privacy concerns. Data stored in the cloud or shared between organizations may be subject to unauthorized access, hacking attempts, or even misuse by service providers. This can result in a loss of control over personal data and raise questions about the transparency and security of AI applications.

To address these privacy concerns, there is a growing emphasis on the development and implementation of privacy-preserving AI techniques. These techniques aim to protect sensitive data while still enabling AI systems to learn and make accurate predictions. Examples include federated learning, which allows data to remain on users' devices while models are trained collaboratively, and differential privacy, which adds noise to data to protect individual privacy while maintaining overall data utility.

Regulatory frameworks and privacy laws also play a crucial role in safeguarding individual privacy in the context of AI. Governments and organizations are increasingly recognizing the need to establish robust data protection regulations and enforce strict privacy practices. Compliance with regulations such as the European Union's General Data Protection Regulation (GDPR) and the California Consumer Privacy Act (CCPA) helps to ensure that personal data is handled responsibly, and individuals have greater control over their information.

The use of AI technologies raises significant privacy concerns due to the reliance on large datasets and the potential exposure of sensitive information. It is essential for organizations, policymakers, and AI developers to prioritize privacy protection, implement secure data handling practices, and comply with privacy regulations to mitigate the risks and ensure that AI is developed and deployed in a privacy-conscious manner.

The AI-Cybersecurity Skills Gap

The rapid evolution of artificial intelligence (AI) and its increasing application in the field of cybersecurity present both opportunities and challenges. While AI offers powerful tools for detecting and responding to cyber threats, its advancement often outpaces the development of necessary skills in the cybersecurity workforce. This creates a skills gap that can leave organizations vulnerable to emerging AI-driven cyber threats.

The integration of AI technologies into cybersecurity practices requires a highly skilled workforce capable of understanding and effectively utilizing these tools. However, the demand for AI experts with deep knowledge and expertise in cybersecurity often exceeds the supply. The rapid pace of AI development, coupled with the complex nature of cybersecurity threats, makes it challenging for organizations to keep up and acquire the necessary talent.

This skills gap poses several risks to organizations. First, it hinders their ability to fully leverage AI technologies for effective cybersecurity defense. Organizations may struggle to implement and manage AI-powered cybersecurity systems due to a lack of skilled professionals who can develop, configure, and maintain these systems. Without the necessary expertise, organizations may not be able to take full advantage of AI's capabilities to detect and respond to advanced cyber threats.

Second, the shortage of skilled professionals in the cybersecurity workforce creates a talent vacuum that cybercriminals can exploit. Adversarial actors with advanced AI knowledge and skills may be able to exploit vulnerabilities and evade detection by using AI techniques in their attacks. The lack of cybersecurity professionals who understand AI-driven threats can leave organizations ill-equipped to defend against such attacks.

To address this skills gap, organizations and educational institutions need to invest in training and development programs that focus on AI in cybersecurity. This includes providing opportunities for cybersecurity professionals to acquire AI skills through specialized training, certifications, and hands-on experience. Collaboration between academia, industry, and government is crucial to develop comprehensive educational curricula and bridge the gap between AI and cybersecurity expertise.

Furthermore, organizations can foster a culture of continuous learning and innovation within their cybersecurity teams. This involves encouraging professionals to stay updated with the latest developments in AI and cybersecurity, promoting knowledge sharing and collaboration, and providing resources for ongoing professional development.

The rapid evolution of AI in cybersecurity creates a skills gap that organizations must address to effectively defend against AI-driven cyber threats. The shortage of skilled professionals with expertise in both AI and cybersecurity poses challenges for organizations in implementing and managing AI-powered security systems. To mitigate this gap, investing in training and development programs, fostering collaboration, and promoting continuous learning are essential to build a capable cybersecurity workforce that can harness the potential of AI while protecting against emerging threats.

In the face of these challenges, robust cybersecurity measures, regulatory oversight, and a focus on developing the necessary skills in the workforce are essential to harness the benefits of AI while mitigating the risks.

IV. The Balancing Act of Regulation

AI and its applications are rapidly evolving, generating both awe-inspiring innovations and equally substantial risks. As we begin to integrate AI more deeply into our lives, the question arises: How do we strike a balance between encouraging innovation and managing the risks that come with it? This is where regulation enters the picture.

Encouraging Innovation

Artificial intelligence (AI) has emerged as a transformative technology with the potential to revolutionize industries, enhance productivity, and address complex global challenges. Given its far-reaching impact, it is crucial to establish regulatory frameworks that not only facilitate AI research and development but also ensure responsible and ethical use of AI technologies.

Regulation plays a vital role in shaping the development and deployment of AI. By providing clear guidelines and standards for responsible AI usage, regulations can help build public trust and acceptance of AI technologies. When individuals have confidence that AI systems are designed and used in a transparent and accountable manner, they are more likely to embrace these technologies and leverage their potential benefits.

Additionally, regulatory frameworks can create an environment that encourages investment in AI. Intellectual property protections, such as patents, copyrights, and trademarks, incentivize organizations to invest in AI research and development by granting them exclusive rights over their innovations. This protection fosters innovation and rewards the investments made by AI developers, encouraging further advancements in the field.

Furthermore, regulations can provide incentives for research and development activities in AI. Governments can offer tax benefits, grants, or other forms of financial support to organizations engaged in AI R&D. These incentives encourage organizations to invest in cutting-edge AI technologies, foster collaboration between academia and industry, and drive innovation in AI-related fields.

Regulatory frameworks should also prioritize ethical considerations in AI development and deployment. This includes addressing issues such as bias, fairness, privacy, and accountability. Regulations can establish guidelines for data collection, usage, and protection to ensure that AI systems operate in a manner that respects individuals' rights and maintains societal values.

International collaboration is crucial in shaping regulatory frameworks for AI. Organizations such as the Organization for Economic Co-operation and Development (OECD) provide guidance and principles for the responsible development and deployment of AI. These principles emphasize the importance of transparency, accountability, and inclusiveness, providing a foundation for countries to develop their own regulations that align with ethical standards.

Regulatory frameworks play a critical role in shaping the development, deployment, and responsible use of AI technologies. By providing clear guidelines, fostering public trust, and incentivizing investment in AI research and development, regulations can support innovation and ensure the ethical and beneficial use of AI. Collaborative efforts among governments, industry stakeholders, and international organizations are essential to establish effective and adaptable regulatory frameworks that harness the potential of AI while addressing societal concerns.

Managing Risk

While AI holds immense potential for innovation and advancement, unregulated deployment of AI systems can introduce significant risks and societal challenges. These risks include threats to privacy, potential biases in decision-making, job displacement due to automation, and even existential risks associated with the development of highly advanced AI systems. To mitigate these risks, thoughtful regulation is necessary.

Regulatory frameworks play a crucial role in ensuring the responsible and ethical use of AI. They can enforce necessary safety standards to ensure that AI systems operate reliably and do not pose harm to individuals or society. By setting guidelines and requirements for transparency, regulations can promote accountability and trust in AI technologies. This includes the disclosure of the decision-making processes of AI algorithms, allowing users and stakeholders to understand how and why certain decisions are made.

Privacy protection is another essential aspect that regulation can address. AI systems often rely on vast amounts of data, and regulations can establish rules to safeguard individuals' privacy rights. This includes ensuring informed consent for data collection and usage, setting limits on data retention, and preventing unauthorized access or misuse of personal information.

Moreover, regulation can address the social implications of AI, such as job displacement. By implementing policies that support job retraining programs and social safety nets, governments can help individuals transition to new roles and industries as automation affects traditional job markets. This can mitigate the potential negative impacts of AI on employment and ensure a just and inclusive society in the face of technological advancements.

Regulations should also consider the broader societal impact of AI, including its potential biases and discriminatory effects. By requiring transparency in AI algorithms and addressing biases in training data, regulations can promote fairness and equity in AI applications, minimizing the risk of perpetuating existing social inequalities.

Lastly, regulation can also play a role in addressing the risks associated with the development of highly advanced AI systems. AGI systems possess capabilities that surpass human intelligence and could potentially have far-reaching impacts. Regulation can help ensure the safe and responsible development of AGI, implementing measures to prevent unintended consequences and addressing the ethical and existential risks associated with this technology.

While AI has the potential to bring about significant benefits, unregulated deployment can pose risks to individuals, society, and even humanity as a whole. Thoughtful regulation is essential to mitigate these risks, enforce safety standards, protect privacy, address social implications, and ensure fairness and transparency in AI systems. By striking the right balance between fostering innovation and safeguarding against potential harms, regulations can promote the responsible and ethical use of AI, fostering a positive impact on society.

Challenges in AI Regulation

Regulating AI is undoubtedly a complex and challenging task due to several factors. Firstly, the rapid pace of AI development often outpaces the traditional process of lawmaking. By the time regulations are enacted, AI technologies may have advanced significantly, rendering the regulations outdated or unable to keep up with emerging challenges. This creates a gap between the capabilities and potential risks of AI and the regulatory frameworks in place.

Additionally, the technical complexity of AI poses a significant hurdle for policymakers. AI systems utilize intricate algorithms, machine learning models, and data processing methods that can be difficult for non-experts to comprehend fully. Policymakers must understand the underlying technology and its potential implications to craft effective regulations. Without a solid understanding of AI, there is a risk of enacting laws that are either too restrictive or insufficient to address the specific risks and nuances of AI applications.

Another aspect that makes regulating AI challenging is its global nature. AI transcends national borders, with research, development, and deployment taking place on an international scale. This presents difficulties for individual nations trying to regulate AI on their own, as inconsistent regulations across different jurisdictions can create fragmentation and hinder the development and deployment of AI technologies. To effectively regulate AI, international cooperation and coordination are necessary to harmonize legal frameworks and address the global challenges associated with AI.

International cooperation entails aligning various interests, legal systems, and cultural perspectives. Finding common ground among diverse stakeholders, including governments, industry leaders, researchers, and civil society, is a complex and delicate process. It requires extensive dialogue, negotiation, and the establishment of international agreements or standards to ensure consistent and responsible AI practices across borders.

Moreover, regulation must strike the right balance between promoting innovation and ensuring safety, privacy, and ethical considerations. Excessive regulation could stifle AI development and hinder its potential benefits, while inadequate regulation may result in unaddressed risks and negative consequences. Achieving this balance requires ongoing assessment and adjustment of regulatory frameworks to keep pace with the evolving landscape of AI technologies.

Regulating AI presents numerous challenges due to the rapid pace of development, technical complexity, and global nature of AI. Policymakers face the task of crafting regulations that are up to date, comprehensive, and adaptable to future advancements. International cooperation is crucial to harmonize regulations and address global challenges associated with AI. Striking the right balance between promoting innovation and safeguarding against risks requires a thorough understanding of AI technology and its potential impacts on society.

V. Case Study: The European Union's Approach to AI Regulation

The European Union's approach to AI regulation exemplifies a proactive and comprehensive effort to balance innovation and risk mitigation. In 2021, the EU introduced the proposed Artificial Intelligence Act, which aims to establish a robust legislative framework for AI.

The Artificial Intelligence Act is designed to address the potential risks associated with AI while fostering innovation and promoting trustworthy AI systems. One of the key objectives is to provide clear rules and guidelines for the development, deployment, and use of AI technologies. By establishing these rules, the Act aims to enhance transparency, accountability, and user trust in AI systems.

Transparency is a crucial aspect of the proposed regulation. The Act introduces requirements for AI developers and providers to provide clear information about the capabilities, limitations, and intended purposes of their AI systems. This transparency aims to prevent the deployment of AI technologies that may have unintended consequences or biases, ensuring that users have a clear understanding of how AI systems operate and make decisions.

The proposed Act also focuses on addressing specific risks associated with certain high-risk AI systems. High-risk AI systems are those that could potentially impact individuals' fundamental rights or pose significant risks to health, safety, or legal compliance. Examples of high-risk AI systems include those used in critical infrastructures, transportation, healthcare, and law enforcement. The Act imposes stricter requirements and obligations on developers and users of high-risk AI systems to ensure their safety, robustness, and ethical use.

The Artificial Intelligence Act emphasizes the importance of human oversight and control in the deployment of AI systems. It highlights the need for human intervention and explains that AI systems should be designed in a way that enables human users to exercise their own judgement. This provision aims to avoid over-reliance on AI systems and ensure that human values, ethics, and responsibility remain central in decision-making processes.

To enforce compliance with the regulation, the proposed Act introduces substantial penalties for non-compliance, including fines that can reach up to 6% of a company's total global annual turnover. By imposing strict penalties, the Act seeks to incentivize adherence to the established rules and guidelines, encouraging responsible and ethical AI practices.

The European Union's approach to AI regulation, as demonstrated by the proposed Artificial Intelligence Act, underscores the importance of striking a balance between fostering innovation and addressing potential risks. By establishing clear rules, enforcing transparency, and imposing penalties for non-compliance, the EU aims to create a regulatory framework that promotes the development of trustworthy and ethical AI systems.

The Future of AI Regulation

As AI continues to advance, it is crucial for AI regulation to evolve alongside it. The rapid pace of technological development requires agile and adaptable policy frameworks that can effectively address emerging challenges and opportunities in the AI landscape. This means that regulation should not be static but rather dynamic, capable of keeping up with the evolving nature of AI technologies.

To ensure the responsible and ethical use of AI, regulation should be grounded in principles that prioritize human rights, fairness, transparency, and accountability. Ethical considerations should guide the development of AI governance frameworks, ensuring that AI technologies align with societal values and respect individual rights. This involves addressing potential biases, ensuring algorithmic transparency, and safeguarding privacy and data protection.

Regulatory approaches to AI governance can take various forms, including both hard law and soft law mechanisms. Hard law refers to legally binding regulations and legislation imposed by governments or international bodies. Soft law, on the other hand, encompasses industry standards, self-regulatory initiatives, and codes of conduct developed by AI practitioners, companies, and professional organizations. The combination of hard and soft law allows for a more comprehensive and adaptable approach to AI governance.

In addition to legal and industry-based regulations, an ethics-based approach to AI governance is becoming increasingly important. Ethical guidelines and principles can complement regulatory frameworks by providing a moral compass for the development and deployment of AI technologies. Ethical considerations encompass issues such as bias mitigation, algorithmic transparency, fairness, accountability, and the social impact of AI.

Creating a successful regulatory environment for AI requires striking a delicate balance between promoting innovation and managing risks. It is essential to foster an AI ecosystem that encourages responsible and ethical practices while fostering innovation, investment, and competition. Effective regulation should support and incentivize AI research and development, while also ensuring that the benefits of AI are shared equitably and that potential risks are mitigated.

As AI continues to shape various aspects of society, including healthcare, finance, transportation, and governance, the importance of robust and adaptable AI regulation cannot be overstated. It is a complex task that requires collaboration among policymakers, industry stakeholders, AI researchers, ethicists, and civil society. By fostering an inclusive and multidisciplinary dialogue, we can develop regulatory frameworks that strike the right balance between technological advancement, ethical considerations, and societal well-being.

Looking ahead at the evolution of AI regulation, it will be essential to harness the full potential of AI while ensuring that it serves the best interests of humanity. By embracing agile policy frameworks, a blend of hard law, soft law, and ethics-based approaches, we can shape an AI ecosystem that is robust, ethical, and beneficial for all.

7

"Coding is the new literacy. It will be as important for future generations to understand artificial intelligence and how to code it as it is to understand math and science."

- Fei-Fei Li, Co-director of Stanford University's Human-Centered AI Institute.

Preparing for the AI Revolution

The Importance of Coding in the AI Era

In our increasingly digital world, coding is quickly becoming an essential skill. This is especially true in the AI era, where the ability to program computers plays a pivotal role in creating and controlling intelligent machines. But coding isn't just about talking to computers; it's a way of thinking and problem-solving that's becoming increasingly important in a variety of careers and everyday tasks.

I. Understanding Coding

In its simplest form, coding is a way of giving instructions to a computer. Imagine a computer as a chef in a kitchen, and coding as a recipe that guides the chef through the steps of preparing a meal. A computer code, or program, is a set of instructions that a computer follows to perform a particular task. These instructions are written in a programming language, which serves as the language of communication between humans and computers.

Programming languages are designed to be human-readable, allowing programmers to write instructions in a format that is understandable and logical. These instructions are then translated into machine-readable code by a compiler or interpreter, enabling the computer to execute the program and perform the desired task.

When it comes to AI, coding plays a critical role in developing and implementing AI algorithms and models. AI algorithms are a set of mathematical instructions that enable computers to learn from data, make predictions, and perform tasks without explicit programming. Programming languages such as Python, Java, and C++ are commonly used in AI development, providing the necessary tools and libraries to build AI systems.

Coding is often compared to cooking because, like cooking, it requires a combination of technical skills, creativity, and problem-solving abilities. Just as a chef selects the right ingredients, follows a recipe, and adjusts cooking techniques to create a delicious dish, a programmer selects the right programming language, writes the code, and debugs and refines it to create a functional and efficient program.

Moreover, coding in AI involves designing and implementing algorithms that can process large amounts of data, identify patterns, and make intelligent decisions. This requires creativity and problem-solving skills to develop innovative approaches and algorithms that can tackle complex AI tasks.

So, coding is the process of writing instructions for a computer to follow, just like a recipe guides a chef in the kitchen. It is an essential skill in AI development, allowing programmers to design, implement, and optimize algorithms that power AI systems. With the right combination of technical expertise, creativity, and problem-solving abilities, coding enables individuals to unlock the potential of AI and contribute to its advancement.

II. Learning to Code

Learning to code is more accessible than ever, thanks to numerous online platforms that offer coding courses. Whether you're a beginner or looking to enhance your coding skills, these courses provide a structured learning environment to grasp the fundamentals of coding. Many of these online courses are either free or available at a low cost, making them accessible to a wide range of individuals.

Online learning platforms such as Codecademy, Coursera, and Khan Academy offer a variety of coding courses in different programming languages. These courses are designed to guide beginners through the basics of coding, from understanding the syntax of a programming language to writing their first computer program. They often provide interactive coding exercises, quizzes, and projects to reinforce learning and practice coding skills.

When embarking on the coding journey, consistency and practice are key. Coding is a skill that develops with practice, much like learning to play a musical instrument. Regular practice helps to solidify the concepts, improve problem-solving abilities, and build confidence in coding.

It's also beneficial to work on real-world projects as part of the learning process. Starting with small projects, such as building a simple website or creating a basic game, allows learners to apply their coding skills in practical scenarios. This hands-on experience not only reinforces learning but also fosters creativity and problem-solving skills.

Furthermore, online coding communities and forums provide opportunities to connect with other learners and experienced programmers. Participating in coding challenges, discussing coding problems, and seeking feedback from peers can further enhance the learning experience and provide valuable insights.

Learning to code is more accessible than ever, thanks to online platforms that offer coding courses. Taking advantage of these resources, beginners can learn the basics of coding and progress to more complex concepts. By practicing regularly, working on real-world projects, and engaging with coding communities, individuals can develop their coding skills and unlock a world of possibilities in AI and other domains (Vihavainen, 2013).

III.   Coding in the Workforce

Coding is increasingly recognized as a valuable skill in the workforce, and its demand continues to grow. According to a report from Burning Glass Technologies, programming jobs are growing at a rate of 12% faster than the market average. This indicates a strong demand for individuals with coding skills in various industries (Burning Glass Technologies, 2016).

Moreover, coding skills often come with higher earning potential. The same report highlights that jobs requiring coding skills tend to pay up to $22,000 per year more than jobs that don't require coding skills. This premium reflects the value and demand for individuals who possess coding expertise (Burning Glass Technologies, 2016).

While coding skills are particularly sought after in the tech industry, they are becoming increasingly relevant in other fields as well. From marketing and design to science and healthcare, many professions now require some level of coding knowledge. In these domains, coding skills enable professionals to leverage technology effectively, automate tasks, and analyze data more efficiently.

Even if your job doesn't directly involve writing code, understanding how code works can still be beneficial. It allows you to communicate and collaborate more effectively with technical colleagues, as you can grasp the language and concepts they use. Additionally, having a basic understanding of coding helps you better understand the capabilities and limitations of technology, enabling you to make more informed decisions in your role.

Furthermore, coding skills foster problem-solving abilities, logical thinking, and attention to detail. These transferable skills are highly valued in the modern workforce, regardless of the specific field or industry. Coding encourages a structured approach to problem-solving and cultivates a mindset of continuous learning and adaptation.

Coding skills are not only valuable in the tech industry but are increasingly sought after in various other sectors. The demand for coding expertise is driven by the growth of technology and the need for professionals who can leverage and understand its potential. By acquiring coding skills, individuals can enhance their career prospects, increase earning potential, and gain a deeper understanding of the technological landscape that permeates today's workplace.

IV. The Future of Coding and AI

As AI technology continues to advance and become more sophisticated, coding will remain a critical skill for individuals working with AI systems. However, the nature of coding is expected to evolve alongside these advancements. One significant shift we may witness is the adoption of higher-level programming languages that emphasize what a program should do, rather than the low-level details of how it should do it.

Traditionally, coding has involved writing detailed instructions that guide a computer step-by-step through a specific task. This process requires a deep understanding of the underlying programming language, algorithms, and data structures. However, as AI systems become more capable, they can handle increasingly complex tasks on their own. This could lead to a transition toward higher-level languages that abstract away some of the technical complexities and allow programmers to focus more on the desired outcomes and objectives of their programs.

Higher-level programming languages enable developers to specify the goals and intentions of a program in a more intuitive and human-readable way. Instead of writing intricate lines of code that dictate every minute detail, programmers can leverage these languages to express the desired behavior or logic of the AI system at a higher level of abstraction. This shift can make coding more accessible to a broader range of individuals and potentially empower more people to create and utilize AI systems.

By utilizing higher-level programming languages, individuals without extensive coding backgrounds can engage with AI technologies more effectively. These languages allow users to interact with AI systems through user-friendly interfaces and natural language commands, reducing the need for in-depth technical knowledge. This accessibility can encourage greater collaboration and innovation across various disciplines, as individuals from diverse backgrounds can contribute their unique perspectives and expertise to AI development and utilization.

Additionally, the adoption of higher-level programming languages in the AI domain can lead to increased productivity and efficiency. Programmers can focus on the strategic aspects of their code, such as defining the problem, selecting appropriate algorithms, and designing effective models, rather than getting caught up in intricate implementation details. This shift allows developers to work at a higher level of abstraction and can accelerate the development process, enabling faster iterations and experimentation.

As AI continues to advance, coding will remain a vital skill (Bau et al., 2020). However, the nature of coding is likely to evolve, with a shift towards higher-level programming languages that emphasize what a program should do, rather than how it should do it. This transition can make coding more accessible, empower individuals from diverse backgrounds to engage with AI systems, and enhance productivity and collaboration in AI development and utilization.

Embracing Coding in the AI Era

In the AI era, coding has transformed from being just a technical skill into a form of literacy that empowers individuals to engage with technology in new and powerful ways. Just as reading and writing enable us to communicate and share ideas, coding allows us to understand and interact with the digital world that surrounds us. By embracing coding, we can all participate in shaping the future of AI and harness its potential for positive impact.

Coding is a means of communication with computers, much like speaking a language or writing a story. It allows us to give instructions to machines, guiding them to perform specific tasks or solve complex problems. Through coding, we can create software, develop applications, and build AI systems that improve our lives and address real-world challenges.

By learning to code, we gain the ability to understand and shape the technology that shapes our world. We become active participants rather than passive consumers, and we gain the skills to transform our ideas into tangible digital solutions. Coding empowers us to take control, to innovate, and to bring our visions to life.

Moreover, coding is not limited to a select group of individuals or professionals. It is a skill that can be learned and utilized by people from all walks of life. From students and entrepreneurs to professionals and retirees, coding provides an avenue for everyone to engage with technology and contribute to the AI revolution.

When we embrace coding as a form of literacy, we open up a world of possibilities. We become problem solvers, equipped with the tools to address challenges, and create solutions that have a meaningful impact on our lives and society as a whole. Whether it's developing an app that simplifies daily tasks, designing a website that showcases creativity, or building an AI system that tackles complex problems, coding empowers us to make a difference.

Furthermore, coding is not just about the end result; it's about the journey of learning and discovery. It cultivates important skills such as critical thinking, logical reasoning, and creativity. Coding challenges us to break down problems into smaller, manageable parts and to think systematically to find solutions. It encourages us to experiment, iterate, and embrace failure as a steppingstone to success.

In the AI era, coding is not just for programmers or computer scientists. It is a skill that can benefit individuals in any field or profession. Whether you're an artist, a doctor, a marketer, or an educator, coding can enhance your capabilities and enable you to leverage AI technologies in your work. It allows you to understand the potential and limitations of AI, collaborate effectively with technical experts, and explore new possibilities for innovation and problem-solving.

In conclusion, in the AI era, coding has become a form of literacy that empowers individuals to engage with technology and shape the future. It goes beyond being a technical skill; it is a means of communication, problem-solving, and creativity. By embracing coding, we can all play a part in harnessing the power of AI and driving positive change in our world.

The Emergence of Data Science

As we continue to digitize and connect more of our world, an unprecedented amount of data is being generated every second of every day. This explosion of data has given rise to data science, a field that combines programming, statistical analysis, and domain knowledge to extract meaningful insights from large datasets. Data science is a crucial cog in the AI machine, providing the raw material - data - that AI systems need to learn and improve.

I. What is Data Science?

Data science is a dynamic and interdisciplinary field that leverages scientific methods and advanced technologies to extract valuable knowledge and insights from vast amounts of data. It combines techniques from mathematics, statistics, computer science, and information science to analyze, interpret, and derive meaningful patterns and trends from complex datasets. The ultimate goal of data science is to use this understanding to drive informed decision-making and take actions that lead to positive outcomes (Provost & Fawcett, 2013).

To better understand data science, think of it as a process of asking and answering questions using data. A data scientist begins by formulating a question or problem, such as "What products are most popular with our customers?" They then embark on a journey to explore and analyze relevant data to refine and ultimately answer that question.

The process of data science often involves several key steps. Firstly, data scientists gather relevant data from various sources, such as customer transactions, website interactions, or social media activity. They then clean and preprocess the data to ensure its quality and usability. This may involve removing irrelevant or incomplete data, addressing missing values, and transforming the data into a suitable format for analysis.

Once the data is prepared, data scientists employ a combination of programming, statistical analysis, and domain knowledge to uncover meaningful insights. They use statistical techniques to identify patterns, correlations, and anomalies within the data. For example, they may apply regression analysis to understand the relationship between product sales and various factors such as price, marketing campaigns, or customer demographics.

Data scientists also utilize machine learning algorithms to build predictive models that can make accurate predictions or classifications based on the available data. These models learn from historical data and can be used to make informed decisions or forecasts. For instance, they may develop a recommendation system that suggests products to customers based on their browsing and purchase history.

It's important to note that data science is not solely focused on analyzing data. Its true value lies in utilizing the insights derived from data to inform decision-making and drive action. Data scientists collaborate with stakeholders and domain experts to interpret the results and translate them into practical recommendations or strategies. They communicate their findings through visualizations, reports, and presentations, ensuring that the insights are understandable and actionable to non-technical audiences.

So, data science is a multidisciplinary field that harnesses scientific methods, computational tools, and domain expertise to unlock the potential of data. By combining programming, statistics, and business acumen, data scientists uncover meaningful patterns and insights that drive informed decision-making. By leveraging data science techniques, organizations can gain a competitive advantage, optimize operations, and make data-driven decisions that lead to improved outcomes.

II. Key Skills in Data Science

Data science is a multidisciplinary field that requires a diverse set of skills from various disciplines. Let's explore some of the key skills that play a crucial role in data science and how they contribute to the overall practice, you may have seen some of these before.

Mathematics and Statistics: Mathematics forms the foundation of data science, providing the tools and techniques necessary for analyzing and interpreting data. Statistics help in understanding the patterns and relationships within the data and making reliable predictions. Skills such as probability theory, linear algebra, and calculus are fundamental in data science.

Programming and Computer Science: Programming is an essential skill in data science, as it allows data scientists to manipulate and analyze large datasets efficiently. Python and R are popular programming languages used in data science due to their rich libraries and tools specifically designed for data analysis and modeling. Additionally, knowledge of algorithms and data structures from computer science helps optimize data processing and analysis tasks.

Domain Knowledge: Data scientists need to possess domain-specific knowledge to understand the context and nuances of the data they are working with. This expertise allows them to formulate relevant questions, identify meaningful variables, and interpret the results accurately. For example, a data scientist working in healthcare needs knowledge of medical terminology and practices to analyze healthcare data effectively.

Data Visualization: Communicating insights effectively is a crucial aspect of data science. Data visualization skills involve creating visual representations of data, such as charts, graphs, and dashboards, to convey complex information in a clear and concise manner. Data scientists use visualization techniques to explore data patterns, identify trends, and communicate their findings to stakeholders.

Machine Learning: Machine learning is a subfield of data science that focuses on developing algorithms and models that can learn from data and make predictions or take actions without explicit programming. It enables data scientists to uncover patterns, build predictive models, and automate decision-making processes. Understanding various machine learning algorithms and techniques is essential for building accurate and reliable models.

Communication and Storytelling: Data scientists must have strong communication skills to effectively convey their findings to both technical and non-technical audiences. They need to translate complex data-driven insights into meaningful and actionable insights that stakeholders can understand and act upon. Storytelling skills help data scientists present their findings in a compelling and persuasive manner, ensuring that the insights are understood and utilized effectively.

By combining these skills, data scientists can extract valuable knowledge and insights from data, identify trends and patterns, make predictions, and drive data-informed decision-making (Donoho, 2017).

III. Data Science in Action

Data science is a versatile field that can be applied to virtually any industry or domain that generates or utilizes data. Let's explore some examples of how data science is being applied in different fields and industries:

Business: Data science is revolutionizing the business landscape by enabling companies to harness the power of data for various purposes. It helps businesses understand their customers better, optimize their operations, and make data-driven decisions. For instance, Amazon's recommendation system uses data science algorithms to analyze customer browsing and purchasing patterns, allowing them to suggest relevant products that customers are likely to be interested in (Smith, 2016).

Healthcare: Data science has significant implications for healthcare, ranging from disease prediction and outbreak monitoring to personalized medicine. By analyzing large datasets, data scientists can identify patterns and risk factors for diseases, predict disease outbreaks, and improve treatment strategies. During the COVID-19 pandemic, data science techniques were crucial in modeling and tracking the spread of the virus, guiding public health interventions and resource allocation (Zhou et al., 2020).

Journalism: Data science is transforming journalism by enabling data-driven investigative reporting. Journalists can analyze large datasets to uncover hidden patterns, trends, and insights that can inform their stories. Data journalism is particularly useful for investigating complex social, economic, and political issues, where data analysis provides a factual basis for reporting and enhances transparency and accountability.

Sports: Data science is revolutionizing the world of sports by providing valuable insights for performance analysis and strategy development. Sports teams use data science techniques to analyze player performance, identify strengths and weaknesses, and make informed decisions on player recruitment and game strategies. For example, in basketball, teams use advanced analytics to analyze player movement, shot selection, and defensive strategies to gain a competitive edge (Blei & Smyth, 2017).

These examples highlight how data science is making a significant impact across various industries, from e-commerce and healthcare to journalism and sports. By leveraging the power of data and applying data science techniques, organizations can gain valuable insights, optimize their processes, and make informed decisions to stay competitive and achieve their goals.

IV. Learning Data Science

Data science is an interdisciplinary field that offers multiple paths for individuals interested in becoming data scientists. While formal education programs in data science are increasingly available at universities, there are also alternative routes to acquiring the necessary skills.

Formal Education: Many universities now offer degree programs in data science, providing a structured curriculum that covers various aspects of the field. These programs typically include courses in mathematics, statistics, computer science, and domain-specific knowledge. By pursuing a degree in data science, students gain a comprehensive understanding of the field's fundamental principles and techniques.

Learning on the Job: Data scientists often come from diverse backgrounds such as computer science, statistics, economics, or other related fields. They acquire data science skills through hands-on experience and on-the-job training.

Employers may provide opportunities for employees to upskill in data science by working on real-world projects and collaborating with experienced data scientists.

Online Courses: Online learning platforms like Coursera and edX offer a wide range of data science courses, making it accessible to anyone with an internet connection. These courses cover various topics, from programming languages like Python to advanced machine learning techniques. They provide a flexible and self-paced learning environment, allowing individuals to acquire data science skills at their own convenience.

Self-Learning and Open-Source Resources: Self-learners can take advantage of a wealth of resources available for learning data science. Textbooks, online tutorials, and open-source software libraries provide valuable learning materials. For example, the Python libraries Pandas and Scikit-learn are widely used in data science and have extensive documentation and tutorials available online. These resources enable individuals to gain practical experience and deepen their understanding of data science concepts (McKinney, 2011; Pedregosa et al., 2011).

It's important to note that a combination of formal education, practical experience, and continuous learning is often beneficial in becoming a proficient data scientist. The field of data science is evolving rapidly, and staying updated with the latest tools, techniques, and advancements is crucial for success.

V. Future of Data Science

As our capacity to collect and store data continues to increase, the importance of data science will only grow. However, the field is also evolving. New techniques and technologies, like deep learning and big data platforms, are pushing the boundaries of what's possible with data science.

As we have discussed, one significant development in data science is the rise of deep learning. Again, deep learning is a subset of machine learning that utilizes artificial neural networks to analyze and interpret complex patterns in data (LeCun et al., 2015). This technique has revolutionized fields such as image and speech recognition, natural language processing, and autonomous vehicles.

Another advancement shaping the field is the emergence of big data platforms. With the exponential growth of data, traditional data processing tools and techniques are often insufficient. Big data platforms provide the infrastructure and tools necessary to handle massive datasets, allowing data scientists to extract insights and make informed decisions (Chen et al., 2014). These platforms enable organizations to process and analyze vast amounts of data from various sources, uncovering hidden patterns and trends that were previously inaccessible.

While the opportunities presented by data science are immense, it's important to address the ethical considerations associated with its practice. As data science becomes more prevalent, issues of data privacy, fairness, and bias come to the forefront. Data scientists must be mindful of the potential biases present in datasets, as biased data can lead to unfair or discriminatory outcomes (Dwork et al., 2012). It is crucial to develop methods and frameworks to mitigate these biases and ensure that data science is used in a responsible and ethical manner.

For instance, algorithms used in hiring processes may inadvertently perpetuate biases if the training data is biased towards a specific group. It is essential for data scientists to proactively address these biases, both by carefully curating the training data and by incorporating fairness metrics into the development of algorithms (Zemel et al., 2013). This ensures that decisions made based on data science are fair, unbiased, and inclusive.

Furthermore, data privacy is a significant concern. As data scientists work with sensitive and personal information, ensuring the privacy and security of data is paramount. It is necessary to implement robust data protection measures and adhere to privacy regulations to safeguard individuals' information and maintain public trust in data science practices (European Union, 2018).

In the AI era, data science is not just a job – it's a way of thinking about and understanding the world. It offers a set of tools for making sense of the vast amounts of data we generate, helping us to make better decisions and create a better future.

8

"AI will change the world more than anything in the history of mankind. More than electricity."

- Jack Ma, Co-founder of Alibaba Group

The Necessity of Ethics in AI

The rapid development of Artificial Intelligence (AI) has transformed the way we live and work, leading to impressive advancements in a range of fields. However, as AI systems become more integrated into our lives, ethical considerations around their use have become increasingly important. Ethical awareness in AI is about understanding and addressing the potential impacts and harms that AI technologies can have on individuals and society. We touched on this briefly in an earlier chapter, but the importance of ethical considerations with the use of AI cannot be emphasized enough.

I. Understanding AI Ethics

AI Ethics is an important field of study that focuses on the moral considerations and challenges associated with the development, deployment, and use of AI systems. It addresses a range of ethical issues that arise from the use of AI technologies and aims to ensure that these technologies are developed and applied in a responsible and beneficial manner for individuals and society as a whole.

Bias

Bias in AI algorithms is a significant concern within the field of AI ethics. When AI algorithms are trained on large datasets, they have the potential to learn and perpetuate biases that exist within the data. This can lead to discriminatory outcomes and reinforce existing social inequalities. Addressing bias in AI algorithms is crucial to ensure fairness and equal treatment for all individuals (Barocas & Selbst, 2016).

To illustrate the impact of bias in AI algorithms, let's consider an AI system used in the hiring process. If the training data used to build the algorithm is biased towards a specific group, such as favoring candidates from a certain gender or ethnicity, it can result in unfair hiring practices and perpetuate existing inequalities in the job market. For example, if historically there has been a gender bias in certain industries, with fewer women being hired, the AI algorithm might learn and reproduce this bias by favoring male candidates during the hiring process.

Ethical considerations in AI require data scientists and developers to actively identify and mitigate such biases to ensure fair and unbiased outcomes in the hiring process. This may involve carefully selecting and preprocessing training data to minimize biases, monitoring the algorithm's performance for potential biases, and implementing fairness metrics and techniques to ensure equal opportunities for all candidates (Kusner et al., 2017).

By addressing bias in AI algorithms, we can strive for fair and equitable outcomes in various domains where AI is used. This is essential to avoid perpetuating social inequalities and to ensure that AI technologies are used responsibly and ethically.

Privacy

Privacy is a critical concern within the realm of AI ethics. AI systems typically rely on large volumes of data, including personal and sensitive information. Safeguarding individuals' privacy and ensuring the secure handling of their data is crucial to maintain trust in AI technologies (European Union, 2018).

Imagine using a voice-controlled AI assistant like Siri or Alexa. These systems collect and analyze your voice commands to provide personalized responses and recommendations. To ensure privacy, it is essential that these AI assistants handle your voice data securely and do not share it with unauthorized parties.

Ethical guidelines and regulations, such as the General Data Protection Regulation (GDPR) implemented by the European Union, are designed to protect individuals' privacy rights in the context of AI. The GDPR establishes strict rules regarding the collection, storage, and processing of personal data, including provisions for obtaining informed consent, ensuring data security, and giving individuals control over their data.

For example, under the GDPR, companies that deploy AI systems are required to obtain explicit consent from individuals before collecting and processing their personal data. They must also implement appropriate security measures to protect the data from unauthorized access or breaches. Additionally, individuals have the right to access their data, request its deletion, and object to automated decision-making processes that significantly affect them.

By adhering to ethical guidelines and regulations like the GDPR, organizations and developers can demonstrate their commitment to respecting individuals' privacy rights. This fosters trust and confidence in AI technologies, assuring users that their personal data is handled responsibly and securely.

Transparency and Explainability

Transparency and explainability are critical aspects of AI ethics. As AI systems become more sophisticated and autonomous, it becomes increasingly important to understand how they make decisions. The ability to interpret and explain the algorithms and outcomes of AI systems is crucial for accountability, trust, and the detection of potential biases or errors (Weller et al., 2019).

Think about using an AI-powered credit scoring system. When you apply for a loan or credit card, the lender may use an AI algorithm to assess your creditworthiness. It is essential to understand how the AI system arrived at its decision and what factors were considered. This transparency helps ensure fairness, enables individuals to challenge decisions if needed, and allows regulators to assess the system's compliance with legal and ethical standards.

Ethical considerations require that AI systems are designed to be transparent and provide explanations for their decisions and actions. This includes making the decision-making process understandable to both technical experts and non-experts. Transparent AI systems allow individuals to know why a particular decision was made and to identify potential biases, errors, or unethical practices.

For example, in the healthcare field, imagine a situation where an AI system is used to diagnose diseases based on medical images. It is crucial for the system to provide an explanation for its diagnosis, highlighting the specific features or patterns that led to the decision. This explanation can help doctors understand and validate the AI's recommendations, leading to better patient care and reducing the risks of misdiagnosis.

Researchers and practitioners are actively working on developing techniques and frameworks for explainable AI. These approaches aim to provide understandable explanations for AI systems' decisions, considering factors such as model interpretability, feature importance, and rule extraction.

By incorporating transparency and explainability into AI systems, we can enhance trust, ensure accountability, and detect and mitigate potential biases or errors. This transparency empowers individuals, regulators, and society as a whole to make informed decisions about the use and impact of AI technologies.

Jobs And Employment

Additionally, AI ethics encompasses considerations of the impact of AI on jobs and employment. As AI technologies advance, there is a concern that they may lead to the displacement of human workers. Ethical approaches to AI involve addressing these concerns by promoting the responsible use of AI to augment human capabilities rather than replacing humans altogether (Brynjolfsson & McAfee, 2014).

The idea is not to view AI as a threat to jobs, but rather as a tool that can enhance human productivity and create new opportunities. For instance, in industries like manufacturing, AI can automate repetitive tasks, allowing workers to focus on more complex and creative aspects of their jobs. This can lead to increased job satisfaction and productivity.

To ensure a just transition to an AI-driven workforce, ethical considerations include retraining and upskilling workers to adapt to the changing demands of the job market. By providing training programs and educational opportunities, individuals can acquire the necessary skills to work alongside AI systems. This can involve learning to collaborate effectively with AI technologies, understanding how to interpret and use AI-generated insights, and developing new roles that leverage human strengths like creativity, empathy, and critical thinking.

Moreover, ethical approaches to AI also involve creating new job opportunities that arise from the development and deployment of AI technologies. For example, the growing field of AI research and development itself creates employment opportunities for scientists, engineers, and data analysts. Additionally, AI technologies can stimulate the emergence of new industries and services, leading to job growth in sectors that leverage AI capabilities.

To navigate the complex ethical challenges posed by AI, various organizations and institutions have developed guidelines and frameworks. These resources provide a foundation for ethical AI development and deployment. For example, the Institute of Electrical and Electronics Engineers (IEEE) has published the Ethically Aligned Design, which provides a set of guidelines for developers and practitioners to consider ethical implications throughout the AI lifecycle. The Partnership on AI, a collaborative platform of technology companies, has also published ethical principles to guide AI development and deployment, ensuring that AI is used in ways that align with societal values (Partnership on AI, n.d.).

These guidelines and frameworks help inform the responsible development and use of AI technologies, promoting ethical considerations in decision-making processes. By adhering to these principles, organizations can mitigate potential harms, promote transparency, and ensure that AI technologies are developed and used in a manner that benefits society as a whole.

AI ethics is a field of study that aims to address the moral and societal implications of AI technologies. It involves considerations of bias, privacy, transparency, fairness, and the impact on jobs. By integrating ethical principles into the development and use of AI systems, we can ensure that AI technologies are used responsibly and contribute to the greater good of individuals and society.

II. AI Ethical Issues in Action: Case Studies

There have been several notable real-world examples that illustrate the ethical issues that can arise from the use of AI. One such case involved an AI system used to predict future criminals. The system was found to be biased against black people, leading to concerns about racial discrimination (Angwin et al., 2016). This case highlights the potential for bias in AI algorithms and the importance of ensuring fairness and equal treatment for all individuals.

Another well-known case involved a chatbot developed by Microsoft called Tay. The chatbot was designed to learn from interactions with users on Twitter and engage in conversations. However, within a short period of time, Tay began making offensive and inappropriate statements, showcasing the potential for AI systems to mimic and amplify negative behavior (Vincent, 2016). This example demonstrates the need for ethical considerations in AI development to prevent the propagation of harmful content or behaviors.

These cases serve as important reminders that without proper safeguards and ethical guidelines, AI systems can produce unintended consequences and perpetuate societal biases. It is crucial to carefully design and test AI systems to identify and address potential ethical issues before deployment.

To mitigate such risks, organizations and researchers are actively working on developing ethical frameworks and guidelines for AI development and deployment. These frameworks aim to ensure that AI systems are developed and used responsibly, considering factors such as fairness, transparency, accountability, and privacy.

By considering the ethical implications throughout the entire lifecycle of AI systems, from data collection and algorithm design to deployment and ongoing monitoring, developers can strive to minimize biases, prevent harmful outcomes, and protect individuals' rights and well-being.

III. Building Ethical Awareness in AI: Education and Advocacy

Given the increasing recognition of the importance of ethics in AI, there is a growing demand for education and awareness in this area. Universities and educational institutions are starting to incorporate AI ethics into their curriculum, offering courses and programs that focus specifically on the ethical implications of AI. These courses aim to equip students with the knowledge and skills to navigate the ethical challenges posed by AI technologies.

For example, the University of California, Berkeley offers a course called "Ethics in AI" that explores the ethical considerations and social impact of AI systems (University of California, Berkeley, n.d.). Similarly, the Massachusetts Institute of Technology (MIT) offers courses on "Ethics of Artificial Intelligence" and "Responsible AI Practices" to help students understand and address the ethical dimensions of AI (Massachusetts Institute of Technology, n.d.).

In addition to formal education, online resources and platforms have emerged to provide accessible information and resources on AI ethics. Websites like EthicsNet, OpenAI's Ethics Resources, and the Future of Life Institute's AI Ethics Reading List offer a wealth of materials, articles, and case studies to help individuals understand and engage with AI ethics (EthicsNet, n.d.; OpenAI, n.d.; Future of Life Institute, n.d.).

Advocacy groups and research organizations also play a vital role in promoting ethical awareness in AI. The AI Now Institute, a multidisciplinary research institute, conducts critical research and provides policy recommendations to address the social implications of AI (AI Now Institute, 2021). The

Partnership on AI, a collaborative platform of technology companies and organizations, aims to advance responsible AI practices and develop ethical guidelines for the industry (Partnership on AI, 2021).

These organizations work towards raising awareness, conducting research, and fostering discussions around AI ethics. They contribute to the development of ethical guidelines, policies, and frameworks that promote responsible and accountable AI practices.

In conclusion, the growing recognition of the importance of ethics in AI has led to an increased focus on education and awareness in this field. Universities are incorporating AI ethics into their curriculum, and online resources are readily available for self-learners. Moreover, advocacy groups and research organizations are actively working to research, advocate, and develop ethical guidelines for the responsible use of AI technologies. These efforts aim to ensure that AI is developed and deployed in a way that is aligned with societal values and promotes positive outcomes for individuals and society as a whole.

IV. The Future of AI Ethics

The importance of AI ethics cannot be overstated, especially as AI continues to advance and become increasingly integrated into various aspects of our lives. As AI technologies become more powerful and pervasive, the potential for ethical challenges and implications also grows. Therefore, it is crucial to develop a deep understanding of AI ethics to navigate these challenges and ensure that AI is developed and used in a way that benefits society as a whole.

AI ethics encompasses a wide range of considerations, including issues such as bias, privacy, transparency, accountability, fairness, and the impact of AI on jobs and society.

By addressing these ethical concerns, we can strive to prevent harm, promote fairness and equality, protect individual rights, and ensure that AI is aligned with societal values and goals.

One of the key reasons why AI ethics is gaining prominence is the potential for unintended consequences. AI systems, when not properly designed or deployed, can perpetuate biases, violate privacy, or cause harm to individuals or communities. For instance, facial recognition technologies have been shown to exhibit racial and gender bias, leading to discriminatory outcomes (Buolamwini & Gebru, 2018). These instances highlight the need for ethical guidelines and practices to mitigate such biases and prevent unfair treatment.

Moreover, as AI becomes more autonomous and capable of making decisions on its own, the need for transparency and accountability becomes paramount. Understanding how AI systems make decisions and being able to explain their actions is essential for building trust and ensuring that AI is used responsibly. This is especially important when AI systems are employed in critical domains such as healthcare, finance, or criminal justice, where the stakes are high and the potential for negative impacts is significant.

Furthermore, the ethical implications of AI extend beyond individual or specific contexts. AI technologies have the potential to shape our societies, economies, and even our fundamental rights. Issues such as job displacement due to automation, algorithmic decision-making in public services, and the concentration of power in the hands of a few companies all raise ethical concerns that need to be addressed.

To fully grasp the concepts of AI ethics, it is important to consider not only the potential risks and challenges but also the opportunities and benefits. AI has the potential to bring about significant advancements and positive societal impacts, such as improving healthcare, addressing climate change, and enhancing productivity. Ethical considerations in AI enable us to harness these benefits while ensuring that they are distributed equitably and in line with societal values.

In conclusion, the importance of AI ethics is undeniable as AI continues to evolve and integrate into various aspects of our lives. Understanding and addressing ethical considerations in AI is essential to prevent harm, promote fairness, protect privacy, and ensure that AI technologies are developed and used in a way that benefits society as a whole. By navigating the ethical challenges and making responsible choices, we can shape the future of AI in a manner that aligns with our values and promotes the well-being of individuals and communities.

9

*"The pace of progress in artificial intelligence is
incredibly fast. Unless you have direct exposure
to groups like DeepMind, you have no idea how
fast—it is growing at a pace close to
exponential." - Elon Musk, CEO of SpaceX, and
Tesla*

Adapting To An AI-Driven World

The Need for AI Education

The rapid evolution and integration of Artificial Intelligence
(AI) technologies into society has changed many aspects of our
lives, including how we work, interact, and learn. With this
dramatic shift comes the need to rethink our approach to
education. Learning about AI and its applications isn't just for
technologists and data scientists anymore – it's becoming a
crucial aspect of general education.

I. Why AI Education is Essential

Artificial Intelligence (AI), at one time seen as a futuristic
concept, is now a fundamental part of our daily lives. AI's
influence ranges from the personalized movie or song
suggestions we receive on Netflix or Spotify, to the digital voice

assistants like Amazon's Alexa or Apple's Siri, which help us manage our schedules, answer queries, and control smart home devices (Russell & Norvig, 2016). This is just the tip of the iceberg, as AI is increasingly used across industries, transforming the ways businesses operate, and redefining our job market.

In simple terms, AI is a broad field of computer science that makes it possible for machines to mimic human intelligence, learning from experiences, adjusting to new inputs, and performing tasks that normally require human intelligence. AI is often used to describe machines (or computers) that exhibit capabilities such as problem-solving, recognizing patterns, understanding language, and making decisions.

Understanding AI is not just for tech geeks. Regardless of your profession or technical background, it's valuable for everyone to have a basic grasp of AI and its implications. Why? For starters, knowing about AI can help us navigate the changing job market. As an example, if you're aware that AI is being used in healthcare to read medical imaging, you might decide to develop skills in AI technology and medical terminology to make yourself more competitive for jobs in that field.

Moreover, being "AI literate" helps us be more informed and discerning consumers. Let's say you're a big fan of comedy movies. You notice that your streaming service keeps recommending horror films. Knowing that these platforms use AI algorithms to suggest movies based on your viewing history, you might decide to "train" the algorithm by rating more comedy movies highly and declining horror movies.

Most importantly, understanding AI allows us to participate in crucial conversations about the ethical and societal implications of these technologies. AI presents complex ethical challenges such as potential bias in AI algorithms, privacy concerns due to data collection, and the impact of AI on job displacement and inequality. As an example, imagine an AI system used for hiring that has been trained on past data and starts to reject female candidates for a traditionally male-dominated role. Being informed about these issues empowers us to push for regulations and practices that prevent such biases.

Fei-Fei Li, a respected AI researcher and professor at Stanford University, emphasized the importance of this understanding when she said, "AI is a new scientific infrastructure for humanity, enabling us to tackle problems that would have been otherwise impossible. But it is also our duty to make sure that AI benefits all of humanity." This underlines the need for us to understand and engage with AI, not just as passive consumers, but as active contributors who can help ensure that AI is used responsibly and ethically to benefit everyone.

Understanding AI isn't just useful, it's essential. As AI continues to permeate our lives and reshape our world, it's crucial that we all have a basic grasp of what AI is, how it works, and its potential implications, so we can navigate our AI-driven world with confidence and integrity.

II.   Integrating AI Education in Schools

As Artificial Intelligence (AI) continues to weave itself into the fabric of our society, there's a growing consensus that we need to start educating our children about it from an early age. A bit like we teach them math or science. A number of countries, including China and the United Kingdom, have already begun incorporating AI into their national curriculums, preparing their younger generations for a future where AI will be a norm rather than an exception (Wang et al., 2019; UK Department for Education, 2019).

Imagine AI education as a subject, much like learning about history or geography, but instead, kids are being taught about things like algorithms, data, and machine learning. It's not as complicated as it sounds! The aim here is to give children an understanding of how AI works, what it's capable of, and, importantly, its impact on society.

One aspect of AI education is teaching kids the basics of how AI works. Remember how in school we learned that plants need sunlight and water to grow through photosynthesis? Similarly, in the context of AI, children can learn the basics of how a computer can be 'trained' to recognize patterns or make decisions.

To make this learning fun and engaging, students can use tools like Scratch, a block-based programming language designed for kids. Scratch allows children to create their own interactive stories, games, and animations while learning the basics of coding. It's a bit like playing with digital Lego blocks but each block is a piece of code! And there's a tool called "Machine Learning for Kids," an extension for Scratch, which lets kids train simple AI models. For instance, they could train a model to recognize when they're making a happy face or a sad face on a webcam – it's learning by doing!

But understanding AI isn't just about the technical stuff. It's also about understanding its societal impacts. This means thinking about how AI affects our everyday lives and the world around us. For example, in a discussion in class, students could examine how recommendation algorithms on platforms like YouTube or TikTok influence what we watch and how this can shape our opinions and behaviors. Or how AI might be used to improve, but also potentially intrude upon, our privacy.

Educating our kids about AI is all about preparing them for a future where AI will be an integral part of their lives – whether that's in their jobs, their hobbies, or even how they interact with the world. As the Chinese proverb says, "The best time to plant a tree was 20 years ago. The second-best time is now." The same can be said for AI education. The sooner we start, the better prepared our children will be for the AI-driven world of the future.

III. Higher Education and AI

The field of higher education has witnessed a surge in the popularity of AI-related courses and programs. Nowadays, universities are offering more than just computer science degrees with a focus on AI. They're developing interdisciplinary programs that merge AI with other fields like ethics, policy, or business. Think of it as studying two subjects in one, allowing students to apply AI in various real-world contexts.

Let's take an example of AI combined with business. A student studying this would learn how to develop and use AI technologies, and also how to implement these technologies in a business setting. They might learn about AI's potential in automating tasks, improving decision-making with data analysis, and how to assess the financial implications of integrating AI into a business model. By the end of their studies, they'd have a solid understanding of both AI and business, and how to intertwine the two.

Even in areas like ethics and policy, combining these with AI is becoming increasingly important. Students in these programs might grapple with questions like: What are the ethical implications of AI-driven decision making? How can policies be designed to ensure AI technologies are used responsibly and don't infringe upon people's rights?

But university education isn't the only pathway to learn about AI. The digital age has brought us Massive Open Online Courses (MOOCs), which are free online courses available for anyone to enroll. MOOCs have been a game changer in making education accessible to people worldwide, regardless of their location or financial status. This means that anyone with an internet connection can access high-quality AI education from prestigious universities.

One great example of this is Stanford University's "AI for Everyone" course on Coursera, designed by AI expert Andrew Ng (Ng, 2019). This course is crafted specifically for non-technical learners, so you don't need a background in computer science or maths to understand it. Just like its name suggests, it's for everyone! You'll learn the basics of AI, understand how AI is affecting the world, and even get some tips on how to navigate your business in an AI-driven world.

So, whether you're interested in a traditional university program or prefer to learn from the comfort of your own home through a MOOC, there's an AI educational pathway that can fit your needs. This accessibility and variety of learning opportunities is equipping more and more people with the knowledge and skills needed to navigate, contribute to, and thrive in our increasingly AI-infused world.

IV. Lifelong Learning and AI

Learning about Artificial Intelligence (AI) is very much like a journey rather than a destination. With the fast-paced advancement of AI technologies, our understanding of it has to evolve continuously. This is where adult learning programs and online platforms step in, offering opportunities for continued AI education.

Imagine an online platform as a virtual university. Websites like Coursera, edX, and Udemy are just like that, hosting a variety of courses on AI and machine learning that cater to different skill levels. You could be a complete novice interested in understanding the basics of AI, a professional looking to upgrade your skills, or an enthusiast wanting to delve deeper into the subject - there's something for everyone.

Coursera, for instance, offers "AI For Everyone" (as mentioned earlier) that is great for beginners, but they also have more advanced courses like "Deep Learning Specialization" for those who want to dive deeper. On the other hand, edX provides a Professional Certificate in AI and machine learning, and Udemy hosts a wide range of courses varying from Python for Data Science and Machine Learning to a Masterclass in AI.

But online learning platforms aren't the only avenue for ongoing AI education. Companies too, are stepping up to the plate. More and more businesses are recognizing the value of AI knowledge across a variety of job roles and are investing in training their employees in this domain. For instance, a marketing executive might be trained in AI tools that can analyze customer behavior, while a human resource professional might be educated about AI-driven tools for resume screening and candidate shortlisting.

These companies are catching onto the fact that, regardless of the industry or role, AI knowledge is becoming an invaluable asset. By providing their employees with AI training, these companies not only upgrade their workforce skills but also increase their competitiveness in an increasingly AI-driven market.

So, whether it's through a course on an online platform or training provided by your employer, there are various opportunities to embark on the journey of lifelong learning about AI. Remember, in the words of Albert Einstein, "Once you stop learning, you start dying." With the continuous evolution of AI, it's more important than ever that we keep learning, adapting, and evolving alongside it.

V.   The Challenge of AI Education

The necessity for Artificial Intelligence (AI) education in our modern world is clear. But much like any other journey, there are bumps along the road. We've got a few key challenges we need to address before AI education can become a reality for everyone.

Firstly, think of a teacher in a classroom. The shift to AI education means that they now have to teach a subject that might be entirely new to them. So, they need to be trained in both understanding AI and in how to effectively teach it to their students. It's like expecting a history teacher to suddenly start teaching advanced calculus - they'll need some preparation and training first!

Additionally, creating the right teaching materials and resources is another challenge. This is not just about textbooks or learning modules; it also involves creating hands-on, interactive learning experiences, which are critical for understanding a dynamic field like AI. Picture a science lab in school, with all the equipment for experiments – we need the AI education equivalent of that.

But it doesn't stop there. AI is a rapidly evolving field, with new advancements and discoveries happening all the time. This means the curriculum needs to be regularly updated to stay relevant. It's a bit like updating your smartphone's software to ensure it has the latest features and runs smoothly.

Perhaps one of the most significant challenges is making sure AI education is inclusive and accessible to everyone, no matter where they live or what their financial situation is. This is a big issue because not everyone has the same access to technology or internet connectivity - this is often referred to as the "digital divide."

Think of it like this: Imagine two kids, one in a bustling city with high-speed internet and the latest tech at her fingertips, and another in a remote village where internet access is patchy at best. If we don't address this divide, the city kid will have a massive advantage in learning about AI. That's why it's crucial that we put efforts into providing equal opportunities for AI education, to ensure every kid gets a fair shot at learning these important skills. This might involve providing affordable internet access, investing in tech infrastructure in underserved areas, or offering scholarships for AI courses to students from low-income backgrounds.

In essence, while the road to widespread AI education has its challenges, by acknowledging these obstacles and actively working towards solutions, we can pave the way for AI education to be a reality for everyone. And as we move further into a world where AI becomes more and more embedded in our lives, this effort becomes increasingly important.

VI. The Future of AI Education

As we peek into the future, one thing becomes clear: AI education is here to stay and will continue to be a priority. As AI technologies keep changing and evolving, so will our understanding of what we should be teaching about AI.

Think about it like this: AI is like a fast-growing tree, sprouting new branches all the time. And as the tree grows, we need to constantly update our knowledge and keep learning about these new branches.

One significant shift we're likely to see in the future is a bigger focus on interdisciplinary AI education. This means that instead of just teaching the nuts and bolts of AI, we'll be combining this technical knowledge with a deep understanding of how AI impacts society, ethics, and policy. It's like a mash-up of computer science with sociology, philosophy, and political science. We'll learn not just how to create AI technologies, but also how to ensure these technologies are used responsibly and ethically.

Let's say you're learning about facial recognition AI systems. An interdisciplinary approach would involve not just understanding how the AI identifies faces, but also considering the ethical implications of these systems - like privacy concerns or potential bias in the system.

Moreover, there's likely to be a greater emphasis on cultivating critical thinking and problem-solving skills. In an AI-driven world, these skills are more important than ever. We need to be able to ask the right questions, analyze situations, and solve problems effectively.

For example, if you're working in a company that uses AI to sort resumes, critical thinking would involve asking questions like: Is the AI accidentally filtering out qualified candidates because of bias in its programming? If so, how can we solve this problem?

The future of AI education is not just about training more programmers or engineers, but also about nurturing thoughtful, informed citizens who can navigate and shape the AI-driven world. This way, we're not just passive consumers of AI technology, but active participants, able to use AI responsibly and to its fullest potential.

The AI Job Market

Artificial Intelligence (AI) is transforming the global job market, redefining roles, and creating new opportunities. While it's natural to fear job displacement due to automation, the AI-driven world also opens up many exciting possibilities. Here's a look at how the job market is changing and how to adapt to it.

I. Job Transformation and AI

There's a lot of chatter about how Artificial Intelligence (AI) is a job killer. But here's the twist: AI doesn't just eliminate jobs;

it changes them, often for the better. Many of the roles we have today actually include an AI component (Bessen, 2019).

Consider a marketing professional's job. Back in the day, they'd probably be spending a good chunk of their time analyzing customer data, running surveys, and manually tweaking their campaigns based on feedback. Nowadays, many of these professionals are using AI algorithms to do a lot of this heavy lifting. These algorithms can predict customer behavior and tailor campaigns to individual customers, making the whole process more efficient and personalized. In other words, the role of a marketing professional hasn't been eliminated; it's been transformed to incorporate AI tools.

But the influence of AI doesn't stop there. It's not just changing the kinds of tasks we do in our jobs; it's also changing how we do our work. AI-powered tools can take over many of the mundane and routine tasks that take up our time. This automation allows professionals to focus on the more strategic, creative, and complex aspects of their jobs, essentially improving the quality of our work lives.

Take doctors, for example. They often have to juggle vast amounts of patient data - everything from medical histories to test results. But with AI, much of this data analysis can be automated. An AI system can quickly sift through the data, spot trends, and even make predictions, freeing up doctors to spend more time on patient interaction and reviewing complex cases (Obermeyer et al., 2019). So, instead of replacing doctors, AI is helping them focus more on the human aspects of their job, which are often the most rewarding.

To sum it up, AI isn't the job-killing monster it's often made out to be. Instead, it's more like a powerful tool that's transforming our jobs and how we work, often in ways that let us focus on what we humans do best: think creatively, solve complex problems, and connect with each other on a human level.

II. New Job Opportunities and AI

Just like a game of Tetris, as some old job roles disappear, new ones pop up in their place. While Artificial Intelligence (AI) might automate certain tasks, it's also a job creator. It's like a power drill; sure, it might replace the manual effort of turning a screwdriver, but it also opens up new possibilities, like constructing a skyscraper!

Now let's consider some of the new jobs that have sprung up thanks to AI. Roles such as AI specialist, data scientist, machine learning engineer, and AI ethics officer were virtually unheard of a few decades ago. They've emerged as we've found new ways to use AI and as we've realized that we need people to guide its responsible use.

Think about what an AI specialist does. They work on developing, maintaining, and strategizing AI systems. This might involve programming AI algorithms, training AI models using data, or troubleshooting when things go wrong. It's a job that requires a mix of coding skills, a good understanding of AI principles, and problem-solving capabilities.

Data scientists, on the other hand, spend their time diving into the sea of data that's available these days, looking for meaningful insights and trends. They use AI tools to help them sift through this data, and their work can inform everything from business strategies to scientific research.

Then we have machine learning engineers, who are like the architects of AI. They design and build systems that can learn from and make decisions or predictions based on data.

Finally, let's look at the role of an AI ethics officer. This is someone who ensures that the AI technologies a company develops or uses are designed and implemented in a way that respects ethical standards and societal norms. This includes considering aspects like fairness, transparency, and privacy.

According to LinkedIn's 2020 Emerging Jobs Report, AI specialist was the top emerging job, with a 74% annual growth rate over the previous four years (LinkedIn, 2020). That's a significant surge and shows just how in-demand these skills have become.

In essence, while AI is reshaping the job landscape and automating some tasks, it's also opening up entirely new career paths. So, rather than fearing AI as a job destroyer, we should embrace it as a job creator.

III. AI Skills in Demand

AI is no longer a niche topic just for tech geeks. Its growing influence means that understanding AI - what we often call AI literacy - is becoming a must-have skill in many different fields, and not just in the tech industry. It's like being computer literate in today's world - it gives you an edge in many different jobs and situations.

When we talk about AI literacy, we're not just talking about knowing how to code an AI program. We're talking about understanding how AI systems work and what they can do, and also understanding the bigger picture, like the ethical issues they can raise. This kind of understanding can give you a significant advantage in the job market because it's a skill that's in high demand but in short supply.

In terms of technical AI skills, there's a long list that are highly sought after. Machine learning - that's the science of getting computers to learn and make decisions like humans do - is a big one. So is natural language processing, which is all about

THE AI REVOLUTION

getting computers to understand and generate human language. Robotics, which often involves AI, is another key skill.

And then there are the skills that are closely linked to AI, like data analysis (turning raw data into useful insights), coding (writing the instructions for a computer program), and statistical programming (using statistical methods to analyze data and make predictions). According to the World Economic Forum, these are some of the skills that employers are crying out for in 2020 (World Economic Forum, 2020).

But here's something you might find surprising: It's not all about tech skills. As we move further into the AI era, "human" skills are becoming more and more valuable. These are the skills that are hard for AI to replicate, like creativity, complex problem-solving, emotional intelligence, and leadership. These skills are like the perfect dance partners for AI: They complement AI's strengths and fill in for its weaknesses.

So, if you're creative, you might come up with innovative ways to use AI. If you're good at problem-solving, you could find solutions to AI challenges. Emotional intelligence could help you understand how people might react to an AI system, and leadership skills could be crucial in guiding a team to implement AI responsibly.

In short, while AI is reshaping the skills landscape, it's not rendering us humans obsolete. Instead, it's creating a job market where technical AI skills and uniquely human skills are both in high demand.

IV. Preparing for AI Careers

In this rapidly changing world where AI is becoming a driving force in many areas, you might be wondering how you can keep up and stay competitive. Just like you'd pack an umbrella for a rainy day, there are several things you can do to prepare yourself for the AI-driven job market. Here's your "AI preparedness" checklist:

Education: It's the first and one of the most important steps. There are plenty of opportunities to learn about AI, from formal university degrees to online courses. Think of it as building your AI toolkit. You might want to focus on areas like AI itself, machine learning (which is all about teaching computers to learn like humans), data science (turning a mountain of data into useful insights), and more. There's a smorgasbord of learning platforms online like Coursera, edX, and Udemy where you can pick up these skills at your own pace.

Continuous Learning: If there's one thing you can count on in the world of AI, it's that it's always changing. So, learning about AI isn't a one-time deal; it's a lifelong journey. Keeping up with the latest trends, tools, and techniques in AI is like keeping your toolkit shiny and sharp. There are many ways to do this: You could follow AI news sites, attend AI conferences or webinars, or join AI communities where you can learn from and share knowledge with others.

Interdisciplinary Knowledge: This is about mixing and matching your AI knowledge with other fields. It's like cooking: You can create interesting and valuable combinations. For instance, if you pair AI with healthcare knowledge, you could end up with a career in health informatics, where you could use AI to help doctors diagnose diseases or predict health trends. Similarly, combining AI with business skills could lead you to use AI in driving business strategies.

Soft Skills: While AI is great at crunching numbers and spotting patterns, there are certain things that it's not so good at - and that's where your uniquely human skills come in. Skills like creativity, communication, problem-solving, and teamwork are all invaluable in the AI-driven world. They're like your secret weapon: They can help you adapt to new situations, work effectively with others, and find solutions to challenging problems.

So, whether you're a student trying to figure out what to study, or a professional wanting to stay competitive, these steps can help you prepare for the wave of AI. By building your AI toolkit, keeping it up-to-date, combining it with knowledge from other fields, and enhancing your soft skills, you can navigate the AI-driven job market with confidence.

V. Navigating AI Job Displacement

While AI opens up a new world of possibilities and opportunities, it's not all sunshine and rainbows. One of the thorny issues that come with AI is job displacement, or in simpler terms, jobs being taken over by machines. Certain sectors like manufacturing, transportation, and customer service are particularly at risk. For instance, manufacturing jobs could be automated with AI-powered robots, self-driving vehicles could disrupt transportation jobs, and AI chatbots could take over customer service roles (Frey & Osborne, 2017).

Imagine you're working in a car manufacturing plant and one day, a robot takes over your job on the assembly line. Or you're a truck driver and suddenly, self-driving trucks are the new norm. That's the kind of situation we're talking about here.

But it's not all doom and gloom. As societies, we can navigate this transition in ways that soften the blow for those who are affected and even turn it into a positive change. Here are a few strategies:

Re-skilling programs: This is all about helping affected workers switch to new careers by learning new skills. Let's say you used to work in a call center that's now using AI chatbots. A re-skilling program could help you learn new skills, like data analysis or digital marketing, that open the door to new job opportunities.

Social safety nets: These are systems that provide support to individuals who lose their jobs due to AI. It could be unemployment benefits, job placement services, or access to affordable retraining programs.

New models of work and income: This could involve creating new kinds of jobs or ways of working that fit in an AI-driven world. It could also mean exploring new ways to ensure people have a stable income, even if their jobs are taken over by AI. For example, some people are discussing the idea of a universal basic income, where everyone gets a certain amount of money to cover their basic needs.

The ultimate goal here isn't to stop AI from progressing. Instead, it's to shape a future where AI serves to augment human work - meaning it helps us do our jobs better, not replace us. We want AI to enhance productivity - making us more efficient and effective - and drive societal prosperity, making our societies better off as a whole. And we want to do this in a way that doesn't leave anyone behind, avoiding displacement and inequality.

Yes, it's a balancing act and a challenging one at that. But by navigating this transition thoughtfully and proactively, we can make the most of the opportunities that AI offers while minimizing the risks.

Living in the Age of AI

From smartphones to smart homes, Artificial Intelligence (AI) has been integrated into our daily lives, changing the way we work, play, and interact. Let's explore how our lifestyles are adapting to this AI-driven world.

I. AI In Everyday Life

AI is everywhere in our daily lives. Let's start by considering your smartphone. If you've ever asked Siri, Alexa, or Google Assistant a question, you've interacted with AI. These virtual assistants use AI to understand what you're saying, find the answer, and then speak it back to you (Smith, 2019).

For example, let's say you ask Siri, "What's the weather like today?" Siri uses AI to recognize your words, understand that you're asking about the weather, figure out what day "today" refers to, determine your location, find the weather for that location, and then give you an answer in natural language. That's AI at work!

The same goes for streaming platforms like Netflix or Spotify. Ever noticed how they seem to know just what you want to watch or listen to next? That's because they use AI algorithms to analyze your past behavior and make recommendations that they think you'll like. If you've ever binge-watched a show on Netflix because it was in your "Recommended for You" section, you've seen AI in action.

AI is also becoming a key part of our homes, making them smarter and more efficient. Take smart thermostats, like the Nest Learning Thermostat, for instance. They learn your routines - when you wake up, when you go to work, when you usually come home - and adjust the temperature accordingly to save energy and keep you comfortable.

A smart refrigerator is another great example. Samsung's Family Hub refrigerator, for instance, can keep track of what's in your fridge, alert you when you're running low on something, and even suggest recipes based on what you have. It uses AI to identify the items in your fridge and to learn from your habits over time.

Then there are smart security systems like Ring's video doorbell or Google's Nest Cam. They use AI to identify when someone's at your door or when there's suspicious activity around your house. Some can even tell the difference between a person, an animal, and a car.

So, as you can see, AI is not just a concept of the future—it has become a part of our present reality. We are increasingly using AI in our everyday lives, often without even realizing it. As AI continues to advance, we can expect it to become even more integrated into our lives and homes.

II. Health and Wellness in the AI Era

Artificial Intelligence is reshaping the way we approach health and wellness, and it's doing so in some pretty impressive ways. From wearable tech that tracks our steps to AI-powered apps that tailor diet advice to our needs, the advancements are really exciting.

Let's start with the devices many of us wear on our wrists. Wearable fitness trackers like Fitbit, Apple Watch, or Garmin watches aren't just cool accessories—they're small AI powerhouses. They gather loads of health data like heart rate, steps taken, and even how well you sleep. They use AI to analyze that data, spot patterns, and provide you with insights. So, if your heart rate spikes when you're at rest, your device might alert you to this potential health issue (Li et al., 2020).

The AI in these devices can also suggest workouts based on your fitness level and goals. Say you've been doing a lot of running and want to mix it up. Your tracker might suggest a high-intensity interval training (HIIT) workout to challenge different muscles and improve your endurance.

But AI's role in health and wellness doesn't stop there. Have you ever tried an app that gives you personalized diet and nutrition advice? These apps, like MyFitnessPal or Lifesum, use AI to analyze your eating habits and suggest changes. If you're consistently missing out on protein, for instance, the app might recommend adding more lean meats or beans to your meals.

Now let's talk about healthcare. The benefits of AI here are huge, particularly in making healthcare more accessible and personalized. Telemedicine platforms, for instance, often use AI to help figure out what's wrong with a patient. When you type in your symptoms, AI is at work determining what health condition you may have. This is known as triage—it's like what a nurse might do when you first arrive at the hospital.

AI also helps doctors diagnose diseases more accurately. Take Google's DeepMind Health, for example. It uses AI to analyze medical images like eye scans, and it can spot signs of eye diseases that can lead to blindness (Rajkomar et al., 2019). This can help doctors catch these conditions earlier when they're easier to treat.

AI is playing a crucial role in revolutionizing health and wellness, making it more personalized, accessible, and efficient. It's a powerful tool in maintaining our health and potentially saving lives.

III.   AI and Social Interactions

Artificial Intelligence, often simply known as AI, is reshaping the way we communicate and interact with each other. On the one hand, it's breaking down barriers and making communication faster and easier. On the other hand, it's raising questions about how it might impact our social interactions and self-reliance.

Think about the language translation tools we use today, such as Google Translate. These tools use AI to understand and translate different languages, helping to break down barriers between people who speak different languages. Picture yourself traveling in a foreign country where you don't speak the local language. Thanks to AI, you can type a sentence into Google Translate and get a translation in real-time. It might not always be perfect, but it's certainly a major help (Hutchins, 2018).

Then there's the predictive text we use every day on our phones and computers. Powered by AI, these tools predict what you're going to type next based on your past typing behavior. This technology helps us type faster, getting our thoughts out more quickly. You might have noticed this when typing a text message or an email – before you finish typing a word, your phone suggests the rest of it.

But while these advancements are exciting, there are concerns about how AI might impact our social interactions. For instance, could over-reliance on AI assistants like Siri or Alexa make us less self-reliant? Imagine if you become so used to asking Alexa for the weather forecast that you forget how to look it up yourself. It's a concern worth considering as these devices become a bigger part of our lives.

AI is also playing a big role in personalizing our online experiences, from the social media posts we see to the news articles we read. But this can create what's known as an "echo chamber," where we're only exposed to content that reinforces our existing beliefs. Eli Pariser coined the term "filter bubble" to describe this phenomenon in his 2011 book, "The Filter Bubble: What the Internet is Hiding from You." This could limit our exposure to different perspectives and affect our understanding of the world.

As AI continues to become more integrated into our lives, it's crucial that we consider these questions and their potential impacts on society. AI can bring about incredible advancements, but it's important to navigate its challenges and potential downsides thoughtfully.

Fear, Misconceptions, and AI

Artificial Intelligence (AI) has undoubtedly revolutionized our world, but its rapid growth and expansion have also sparked fears and misconceptions. It is crucial to tackle these fears head-on and debunk the misconceptions to truly leverage AI's potential benefits.

I. The Fear of AI "Taking Over"

The idea of AI surpassing human intelligence and taking control has been a staple of science fiction for decades, from the rebellious robots of "I, Robot" to the all-knowing supercomputers of "The Matrix". But while these scenarios make for great storytelling, they're far removed from the reality of AI as we know it today.

Right now, the AI we use is what experts call "narrow AI" (also known as "weak AI"). This means that these AI systems are designed to perform a specific task very well, but that's it. They lack the understanding, consciousness, and general problem-solving abilities of humans (Russell & Norvig, 2016).

Consider this example: A chess-playing AI, like IBM's Deep Blue, can beat grandmasters at chess, but it can't understand the concept of the game or even recognize that it's playing a game. It can't apply the strategies it "learns" from chess to any other problem. That's the limit of narrow AI.

As we develop more sophisticated AI systems, we're striving for something closer to "general AI" or "strong AI" - AI that could understand or learn any intellectual task that a human being can. But we're still a long way off from that. More importantly, even as we develop more advanced AI, we're not just blindly pushing for more intelligence without considering the implications.

AI's development is driven by human programming and oversight. Researchers and engineers are working hard to make sure AI systems operate safely and ethically. Many organizations, like OpenAI and the Partnership on AI, have established ethical guidelines to ensure responsible use of AI. These guidelines stress principles like transparency, fairness, and respect for human rights.

So, while the fear of super-intelligent, uncontrollable AI makes for exciting science fiction, the reality is much more mundane (and safer). AI is a tool, created by humans, with built-in safeguards and restrictions to make sure it's used responsibly.

II. The Fear of AI "Spying"

The rise of AI has undeniably led to increased concerns about privacy. Given the large amounts of data that AI systems handle – often personal or sensitive in nature – these concerns are entirely legitimate. For example, a digital assistant like Alexa needs to process our spoken commands, meaning it has access to whatever we say. Similarly, online recommendation systems such as those on Amazon or Netflix operate by analyzing our browsing and purchasing behavior (Zuboff, 2019).

Yet, it's crucial to note that there are multiple mechanisms in place to protect our data. Laws like the European General Data Protection Regulation (GDPR) and the California Consumer Privacy Act (CCPA) impose stringent regulations on how companies collect, store, and use personal data. They grant individuals rights over their data, including the right to access their data, correct inaccuracies, and even demand its deletion (EU GDPR, 2016; CCPA, 2018).

In addition to regulatory protections, there's a growing emphasis on privacy-preserving AI design. Techniques such as differential privacy, federated learning, and homomorphic encryption ensure that AI systems can analyze and learn from data without compromising individual privacy. For instance, differential privacy allows an AI system to learn from a dataset without revealing information about specific individuals within that dataset (Dwork, 2008).

While it's true that AI uses data to function, the primary goal isn't surveillance but to provide personalized and efficient services. Digital assistants use our data to better understand our commands, while recommendation systems analyze our past behavior to suggest products or movies that we might enjoy. It's less about "spying" and more about tailoring the service to the individual's needs.

Nevertheless, the intersection of AI and privacy is a complex area that's constantly evolving. As consumers, we should remain informed about how our data is used and take advantage of the protections and rights granted to us by law. As a society, we need to continually reassess our regulatory frameworks and ethical guidelines to ensure they keep pace with technological advancements.

III. Debunking Misconceptions: AI's Limits

Artificial intelligence, despite its incredible capabilities and advancements, is not without limitations. While AI can analyze vast amounts of data at impressive speeds and accurately perform specific tasks, it does not have the ability to feel or comprehend human emotions. It's also incapable of understanding context in the way humans do, which can result in misinterpretations and mistakes (Marcus & Davis, 2019).

For instance, if you say something sarcastically to an AI assistant like Siri or Alexa, it's unlikely to catch the sarcastic undertone of your statement. Humans, with their innate understanding of social cues and context, can easily discern the intent, but an AI system will likely interpret your words literally.

AI is also not infallible, despite its reputation for accuracy. It is as good (or as flawed) as the data it has been trained on. If the training data is incomplete, skewed, or biased, the AI system may make erroneous decisions or predictions. A famous example is the AI recruiting tool that Amazon once used, which was found to be biased against female candidates. The reason was that the AI system was trained on resumes submitted to the company over a ten-year period, most of which came from male candidates. As a result, the system unintentionally learned to favor male candidates (Dastin, 2018).

Understanding these limitations is important because it helps us set realistic expectations of what AI can and cannot do. While AI can aid us tremendously in many areas, from healthcare diagnostics to driving autonomous cars, it does not possess human-like understanding, empathy, or judgment. As such, human oversight and intervention remain critical in the deployment and management of AI systems.

As AI continues to evolve, researchers are working to address these limitations. For instance, efforts are underway to create more 'emotionally intelligent' AI and to reduce bias in AI systems. But for the foreseeable future, AI is likely to remain a tool – a highly advanced and powerful one, but nonetheless one that supplements, rather than replaces, human capabilities.

IV. Embracing AI: The Path Forward

Embracing AI – or fully accepting and taking advantage of artificial intelligence – requires much more than simply using AI-powered devices or services. It involves a deeper understanding of AI's workings, its applications, impacts, and the responsibilities that come with its use (Russell et al., 2015).

To start with, we need to promote AI literacy. This means teaching people the basics of how AI works – how it learns from data, makes predictions, and improves over time. For example, an important concept to understand is machine learning, where AI systems learn and improve from experience without being explicitly programmed. The concept of neural networks, which are modeled after the human brain and are the foundation of deep learning, is another fundamental aspect of AI.

At the same time, AI literacy isn't just about understanding the technical side. It's also about understanding how to use AI responsibly and ethically. This might involve considerations about privacy – like not using AI tools to analyze personal data without consent – or fairness, like ensuring AI systems don't perpetuate or amplify existing biases. An example of this is the use of AI in hiring, where care needs to be taken to ensure the algorithms do not discriminate against certain groups (Dastin, 2018).

Embracing AI also means fostering a critical mindset. This means not just accepting AI applications at face value but questioning their impact on society. For instance, if an AI tool is used in decision-making, we should question how the decision was made, whether the process was transparent, and whether the outcomes are fair and beneficial. This critical thinking will help society navigate the challenges of AI and mitigate potential harm.

Lastly, embracing AI means recognizing its potential to transform our world for the better and actively working towards this goal. AI has the potential to revolutionize numerous sectors, from healthcare to education to energy. It can help us diagnose diseases earlier, personalize learning, and reduce energy consumption. But realizing this potential requires active effort, innovation, and collaboration across sectors and disciplines.

In conclusion, embracing AI is a multifaceted endeavor. It's about understanding AI, using it responsibly, fostering a critical mindset, and leveraging its potential for societal good. As we navigate the AI era, these are key principles that will guide us towards a future where AI is used effectively, ethically, and for the benefit of all.

10

"AI is the future of all technological advancements; it is the foundation of any modern society."

- Ginni Rometty, Former CEO of IBM (World Economic Forum, 2019).

Case Studies in AI

Successful Implementations

Artificial Intelligence (AI) has begun to infiltrate almost every sector of industry, from healthcare and education to finance and transportation. Various companies are leading the way in successfully integrating AI into their daily operations. Here are a few notable examples.

I. Google

Google has emerged as a pioneer in applying AI to an array of services that many of us use on a daily basis. Let's dissect their varied applications of AI to truly appreciate the scope of their work.

Search and Ads: Google's search engine, the company's flagship product, has integrated AI to deliver increasingly accurate search results. They use an AI model called BERT (Bidirectional Encoder Representations from Transformers) that helps Google Search better understand the intent behind users' search queries, improving the relevancy of the results displayed (Devlin et al.,

2019). In the advertising realm, Google uses AI to help advertisers target their ads more effectively. The AI analyzes user behavior, preferences, and search history to show the most relevant ads, making the advertising process more efficient and personalized.

Gmail and Google Photos: AI is also deeply embedded in Gmail and Google Photos. Gmail uses AI for features like Smart Reply and Smart Compose, which suggest quick responses or auto-complete emails as you type, respectively (Kannan et al., 2016). Google Photos uses AI to recognize faces and objects in photos, which makes it easier for users to search and organize their photo collections. For example, you could search for "beach" in Google Photos and the service will display all of your beach photos, even if you didn't manually tag them as such (Google AI, 2021).

Healthcare Applications: Google's venture into the healthcare domain showcases AI's potential to transform medical diagnosis and treatment. Google Brain, the company's AI research team, developed an AI system capable of detecting diabetic retinopathy and macular degeneration in eye scans (Gulshan et al., 2016). Diabetic retinopathy is a diabetes complication that can cause vision loss, and early detection is crucial to prevent this. Traditionally, this detection process required a specialist, but with this AI model, the detection can be automated, which could make the screening process more efficient and widely accessible.

Google's application of AI extends across various industries and sectors, making it a global leader in AI development and integration. Their endeavors illustrate how AI can streamline processes, personalize user experiences, and even potentially revolutionize healthcare diagnostics.

Google DeepMind

DeepMind, a British AI company acquired by Google in 2014, has been making significant advancements in AI research, notably in areas such as deep learning and reinforcement learning.

Deep Learning and AlphaGo: Perhaps the most publicized achievement of DeepMind was the development of AlphaGo, an AI program that defeated the world champion Go player in 2016. Go is an ancient Chinese board game considered to have more possible moves than there are atoms in the universe, making it a major challenge for AI. AlphaGo utilized deep learning techniques, in which it was trained on thousands of professional human games and then played against itself millions of times to learn new strategies (Silver et al., 2016). This win was a breakthrough moment in AI, showcasing how AI could master complex tasks by learning from experience.

Reinforcement Learning and AlphaStar: DeepMind's success extends to other games as well. AlphaStar, another DeepMind AI system, achieved Grandmaster level at the video game StarCraft II, which is considered one of the most challenging real-time strategy games. AlphaStar used a type of AI known as reinforcement learning, where the AI learns optimal strategies by playing against itself and updating its strategies based on the outcome of each game (Vinyals et al., 2019).

Healthcare: Beyond games, DeepMind is leveraging AI to solve real-world problems, particularly in healthcare. For instance, DeepMind developed an AI system that can predict the complex 3D shapes of proteins, a task that has baffled scientists for decades. This achievement, recognized as a major breakthrough in the 2020 CASP competition (an event that benchmarks the world's best models at this task), could have significant implications for understanding diseases and developing new drugs (Senior et al., 2020).

AI Safety and Ethics: As part of their AI development, DeepMind is committed to addressing AI safety and ethical issues. They have a dedicated team working on AI ethics and have published research on topics such as fairness in machine learning, interpretability of AI systems, and preventing malicious use of AI (DeepMind, 2021).

In essence, Google's DeepMind is at the forefront of AI research, developing groundbreaking AI technologies and striving to address the ethical and societal implications of AI.

II. Amazon

Amazon, the world's largest online marketplace, is a pioneer in integrating AI across its operations. The company uses AI to improve customer experience, streamline operations, and uncover new business opportunities.

Product Recommendations: Have you ever wondered how Amazon suggests products that are so often of interest to you? This is made possible by their AI-powered recommendation engine. It uses machine learning algorithms to analyze your browsing history, previous purchases, and items in your shopping cart, along with data from other customers with similar behaviors, to predict what products you might be interested in buying next (Smith et al., 2017). This system has been remarkably successful, with Amazon stating that 35% of its revenues come from its recommendation engine (McKenzie, 2013).

Fraud Detection: With millions of transactions happening daily, Amazon is a potential target for fraudulent activity. Here, too, Amazon uses AI to protect itself and its customers. Their AI system uses machine learning algorithms to analyze patterns in purchase activity, identifying anomalies and potential fraud. The system can, for instance, detect if a new account is created and immediately used for high-value purchases, which could indicate a stolen credit card (Sahay, 2019).

Alexa and AI: Amazon's virtual assistant, Alexa, is another great example of their use of AI. Alexa uses AI to understand and respond to voice commands from users, enabling it to play music, answer questions, control smart home devices, and much more. The AI underlying Alexa is continuously learning and improving its ability to understand different accents, languages, and command types (Rauschnabel et al., 2018).

Robotics in Warehouses: Amazon also employs AI to streamline its logistics operations. In its fulfillment centers, Amazon uses robotic systems designed to move around products. These robots use AI and computer vision technologies to navigate the warehouse, avoiding obstacles, and optimizing paths to reduce delivery times (Kootstra, 2020).

In essence, Amazon is a prime example of how AI can enhance business operations and customer experience. From suggesting your next favorite product to delivering it to your doorstep, AI is crucial in almost every step of the Amazon customer journey.

III. Facebook

Facebook, one of the world's most popular social media platforms, is deeply rooted in AI, which powers numerous features of the platform. AI is crucial for enhancing the user experience, moderating content, targeting advertisements, and furthering AI research.

Content Moderation: Given the massive amount of content shared on Facebook every day, manual moderation is practically impossible. To help solve this problem, Facebook uses AI. Machine learning algorithms are trained to detect and remove content that violates Facebook's Community Standards, such as hate speech, graphic violence, or explicit content (Schroepfer, 2020). This doesn't mean AI replaces human moderators. Instead, it's a tool that helps human moderators by identifying and prioritizing potential violations, making the process faster and more efficient.

Personalizing News Feeds: Every time you log in to Facebook, you're met with a personalized news feed. This isn't a random assortment of posts; it's carefully curated by Facebook's AI algorithms. These algorithms consider a multitude of factors like your past interactions, the popularity of posts, and even the type of content in posts (text, photo, video, etc.) to predict what you'd most like to see (Gomez-Uribe & Hunt, 2015).

Ad Targeting: Facebook's AI also plays a significant role in its advertising system. Advertisers can choose the demographic and interest categories of the people they want to target, and Facebook's AI ensures their ads reach those users. The AI system also learns which users are most likely to click on or engage with different types of ads, enhancing ad effectiveness (Varian, 2014).

AI Research: Beyond these applications, Facebook is committed to advancing AI technology itself. Their dedicated AI research division, Facebook AI, conducts research in areas such as computer vision, natural language processing, and artificial general intelligence (AGI), the pursuit of AI systems with broad, human-like intelligence. They make much of their research publicly available, contributing to the wider AI community (Facebook AI, 2021).

However, Facebook's use of AI isn't without controversy. Issues have been raised about user privacy, the potential for bias in AI algorithms, and the effectiveness of AI in moderating harmful content. These are challenges that Facebook, and society as a whole, need to grapple with as we navigate the future of AI.

III. Microsoft

Microsoft, a global technology giant, is at the forefront of integrating artificial intelligence (AI) into its products and services. They've woven AI throughout their offerings, from cloud computing with Azure to productivity tools in Office 365, professional networking in LinkedIn, and beyond.

Office 365: Microsoft's productivity suite, Office 365, uses AI to enhance user experience and boost productivity. For instance, in Word, AI-powered Editor provides advanced grammar and style checking, offering suggestions to improve your writing (Microsoft, 2020). PowerPoint has an AI feature called Designer, which suggests professional designs for your slides based on the content you input. Outlook uses AI for spam filtering, categorizing emails, and suggesting responses, while Excel employs AI for data analysis and trend prediction.

Azure: Azure, Microsoft's cloud computing platform, offers various AI services through Azure AI. These services range from machine learning (building predictive models), speech recognition (transcribing spoken language into text), vision (analyzing and recognizing images and videos), and natural language processing (understanding and generating human language) (Microsoft Azure, 2021).

LinkedIn: Microsoft-owned LinkedIn, the world's largest professional network, uses AI in multiple ways. The most noticeable is the 'Jobs You May Be Interested In' section, which uses AI algorithms to match users' profiles with job postings. Other AI uses include personalizing the news feed and suggesting people you may want to connect with.

AI Research and Advancements: Microsoft is dedicated to pushing the boundaries of AI. Its research team, Microsoft Research AI (MSR AI), explores a wide range of AI areas like machine learning, speech and language processing, and human-computer interaction. One notable breakthrough from MSR AI was the development of an AI system that could understand conversational speech as well as humans do, a significant milestone in speech recognition technology (Xiong et al., 2016). In healthcare, Microsoft has projects like Project InnerEye, which uses AI for automated analysis of medical images, potentially aiding in treatment decisions (Microsoft AI, 2021).

AI Ethics: Microsoft has also shown commitment to responsible AI use. They have published principles for ethical AI use, which guide their product development and business practices. These principles stress fairness, inclusiveness, transparency, reliability, and accountability (Smith, 2018).

Microsoft's AI integrations demonstrate the transformative potential of AI across industries, from boosting productivity in the office to enhancing healthcare diagnostics and treatment. It illustrates a future where AI is an integral part of daily life, augmenting human capabilities and driving innovation.

IV. IBM

IBM's Watson is one of the most prominent AI systems globally, widely recognized for its groundbreaking performance on the quiz show Jeopardy! in 2011. Named after IBM's first CEO, Thomas J. Watson, the AI system leverages natural language processing and machine learning to interpret vast amounts of data and answer complex questions. Watson's capabilities have evolved significantly since its television debut, with applications in diverse fields like healthcare, business, and education.

Jeopardy! and Watson's Unique Abilities: In 2011, Watson outperformed human competitors on Jeopardy!, a quiz show that involves complex language understanding and problem-solving. Watson's success demonstrated AI's ability to understand and generate human language, reason, and learn. In a landmark event, Watson used its "ability to understand complex language, determine the correct answer, and confidence in its answer to dominate the human competition" (Ferrucci, 2012). This showcased the potential of AI in real-world, human-centric tasks, where understanding context and nuances in language is vital.

Watson in Healthcare: Post-Jeopardy!, one of Watson's most promising applications has been in healthcare. Watson Health, a dedicated branch of IBM, aims to use AI to address health challenges. In oncology, for example, Watson can assist doctors by analyzing medical records and clinical guidelines to suggest treatment options. Watson's capacity to process and interpret vast amounts of data can support doctors in making informed decisions (Ferrucci et al., 2013). Additionally, Watson has been used in genetic analysis, drug discovery, and patient care management.

Watson in Business: Beyond healthcare, Watson has applications in various business contexts. IBM offers Watson services for customer engagement, assisting in areas like customer service (with Watson Assistant) and personalized marketing (with Watson Marketing Insights). In the field of financial services, Watson aids in risk management and regulatory compliance.

Future Outlook and Ethical Considerations: As Watson's applications continue to grow, so too does the need for considerations around ethical use. IBM has set forth principles for the deployment of AI, emphasizing trust and transparency, augmenting human intelligence, and skills development (Rometty, 2018).

Watson's journey, from a quiz show to contributing in sectors like healthcare and business, illustrates AI's transformative potential. The evolution of Watson signifies a broader trend: as AI becomes more sophisticated, its applications become more diverse, permeating various aspects of life and work.

V. Tesla

Tesla, under the leadership of CEO Elon Musk, is at the forefront of implementing AI in the realm of autonomous vehicles. The company's goal is to produce vehicles that are not just electric but also self-driving, with its Autopilot feature leading the way in achieving this ambition.

Tesla Autopilot and AI: Autopilot, as the name suggests, is an advanced driver-assistance system that leverages AI and machine learning. The system includes features such as automatic steering, lane changing, and parking, alongside more advanced capabilities such as navigating highways from on-ramp to off-ramp (Tesla, 2021). The AI algorithms running Autopilot are trained on vast amounts of data collected from Tesla vehicles on the road. This data-driven approach allows the system to learn and improve continuously.

Neural Networks and Hardware: Tesla's Autopilot system is powered by neural networks that process data collected by the vehicle's eight cameras, twelve ultrasonic sensors, and front-facing radar. The company also developed a dedicated hardware chip, often referred to as "Hardware 3" or "FSD Computer", to efficiently run these neural networks. Musk has claimed that this chip is the "best chip in the world...by a huge margin" (Hollister, 2019).

Full Self-Driving and the Future: Tesla's ultimate goal is Full Self-Driving (FSD), where the car can operate without human intervention under all conditions. As of my knowledge cut-off in September 2021, FSD is still under development and testing, and its use is regulated by various jurisdictions. Musk has expressed confidence that Tesla is close to achieving this goal, but it's important to note that there are still significant technical and regulatory challenges to overcome (Naughton, 2021).

Safety and Ethical Considerations: With the integration of AI into vehicles, there are substantial safety and ethical considerations. Ensuring the reliability of self-driving systems and how they react in emergency or unforeseen situations is critical. Also, questions about liability in case of accidents and the potential impact on jobs in the transportation industry are being discussed. Tesla's approach is under scrutiny in these aspects, as safety regulators investigate accidents involving Autopilot (Shepardson, 2021).

The work Tesla is doing with AI and autonomous driving represents a significant step towards a future where self-driving cars are commonplace. However, it also underscores the complex challenges - both technical and ethical - that come with integrating AI into our daily lives.

VI. Alibaba

Alibaba, one of the world's largest e-commerce companies, is a prime example of how AI can be embedded across diverse business operations to create value. The company uses AI in various facets of its operations, ranging from customer service to logistics, and even urban planning.

Customer Service with Chatbots: Alibaba utilizes AI-powered chatbots for customer service. These chatbots can handle a wide array of customer inquiries, from tracking orders to handling complaints, and can do so 24/7, increasing efficiency and customer satisfaction. For example, during the 2017 Singles Day shopping festival, the company's chatbot, named "AliMe," handled nearly 95% of customer service inquiries (Aliyun, 2017).

AI in Logistics: AI also plays a crucial role in Alibaba's logistics operations. The company's logistics arm, Cainiao Network, uses predictive analytics to optimize routes, manage inventories, and even predict future demand. This use of AI helps reduce costs, increases efficiency, and improves delivery speed, which is a significant competitive advantage in the e-commerce industry (Fannin, 2019).

City Brain Project: Perhaps one of the most ambitious uses of AI by Alibaba is its "City Brain" project. Launched in 2016 in collaboration with the city of Hangzhou, the project uses AI to improve urban life. City Brain utilizes big data and machine learning to optimize traffic lights and routing, leading to reductions in traffic congestion and improvements in emergency response times (Alibaba Cloud, 2021).

For example, City Brain can automatically adjust traffic light timings based on real-time traffic conditions, prioritize emergency vehicles, and detect traffic incidents. It has been reported that since the implementation of City Brain in Hangzhou, the average travel speed has increased by 15%, and emergency vehicles reach their destinations 50% faster (Alibaba Cloud, 2021).

As Alibaba continues to innovate with AI, it highlights how AI can provide value not only within a company but also for society at large. Alibaba's use of AI underlines how tech companies can become agents of positive societal change through thoughtful integration of technology.

VII.   Finance - JPMorgan Chase

JPMorgan Chase, one of the largest financial institutions in the world, has leveraged artificial intelligence (AI) to enhance efficiency, reduce risk, and drive innovation in its operations. One notable example of its AI integration is the Contract Intelligence (COiN) platform.

The COiN platform is an AI-driven tool that uses a combination of image recognition and natural language processing (NLP) to review legal documents. This process, which previously required extensive human labor, can now be accomplished with dramatically increased speed and accuracy, thanks to AI.

The application of AI to legal document review has clear benefits. In the past, the review of commercial loan agreements was a time-consuming task. These documents can be hundreds of pages long, and interpreting their contents requires a high degree of knowledge and precision. Traditionally, this task was performed by a team of lawyers and loan officers, who would spend thousands of hours reading through the documents to extract important information and identify potential risks (Chui et al., 2018).

With the introduction of the COiN platform, this process has been transformed. Using image recognition, the AI can scan and digitize the documents, converting the text into a format that can be processed by the AI. The natural language processing component then comes into play, analyzing the language in the document, extracting relevant information, and identifying any potential issues. In one test, COiN reportedly managed to complete a task in seconds that would have taken human employees 360,000 hours (Chui et al., 2018).

The success of the COiN platform showcases the transformative potential of AI in sectors beyond the traditionally high-tech industries. In finance and law, professions that have long relied on manual labor and human expertise, AI can automate routine tasks, freeing up human workers to focus on higher-level, strategic tasks. It also shows how AI can enhance accuracy and reduce risk, as the AI system can identify potential issues more consistently than human reviewers.

JPMorgan Chase's implementation of the COiN platform is an impressive example of how AI can drive efficiency and reduce risk in a large, complex organization. It serves as a model for other institutions looking to leverage the power of AI in their operations.

VIII. Agriculture - John Deere

John Deere, a renowned American corporation known for its agricultural, construction, and forestry machinery, has harnessed the power of artificial intelligence (AI) to revolutionize farming practices. They have integrated AI technologies into their machinery to optimize numerous agricultural processes, from crop planting and care to harvesting (Stone, 2020).

AI technologies implemented by John Deere aim to enhance farming operations through precision agriculture. This field of study focuses on managing variations in the field accurately and efficiently to grow more crops using fewer resources, leading to higher yields and less environmental impact.

AI-powered planting and care: John Deere's AI-enabled machinery assists in precision planting and crop care. For instance, their ExactEmerge row unit uses machine vision and machine learning technologies to ensure seeds are placed accurately and optimally in the field for maximum yield. As it moves across the field, the AI system adjusts in real time to variations in soil type, moisture levels, and other field conditions. This intelligent seeding process can reduce waste and increase productivity.

AI in Fertilizing: John Deere also applies AI for precision fertilizing. Their technology can make real-time decisions about when and where to apply fertilizers, considering factors such as soil conditions and crop health. By delivering the right amount of fertilizer at the right time and place, this system helps farmers reduce waste and improve the efficiency of their operations.

AI in Harvesting: The company's AI technologies also extend to harvesting. John Deere's Combine Advisor, for instance, is a suite of seven technologies that optimizes the performance of its combines. Utilizing AI, the system automatically adjusts the combine harvester based on the changing conditions of the field to minimize grain loss and maximize harvesting efficiency.

John Deere's use of AI in farming provides a powerful example of the broader trend of digital transformation in the agricultural sector. This 'smart farming' approach has the potential to revolutionize the industry by driving efficiency, reducing environmental impact, and increasing yield.

Overall, the use of AI by John Deere exemplifies how technology can dramatically enhance traditional industries. By applying AI to improve decision-making and automate processes, John Deere is helping farmers to work more efficiently and sustainably.

IX.   Manufacturing - Siemens

Siemens, a multinational conglomerate known for its expertise in various fields, such as automation, digitalization, and electrification, has been a trailblazer in integrating artificial intelligence (AI) into manufacturing. Their AI applications are designed to automate processes, maintain quality control, and predict maintenance needs, consequently improving efficiency and reducing costs (Siemens, 2021).

AI for Process Automation: One area where Siemens is utilizing AI is in process automation. This involves using machine learning algorithms and robotic process automation to take over repetitive tasks, freeing up employees to focus on more complex, value-added activities. This not only increases productivity but also reduces the risk of human errors.

AI in Quality Control: AI has a significant role in ensuring product quality in manufacturing. Siemens has developed a system, which they call "AI Quality Inspector," that uses machine vision to automatically detect defects in the manufacturing process. This system can analyze images of products taken from multiple angles and identify anomalies that might indicate a defect. For example, it might find a dent in a sheet of metal or a missing component on an assembly line. By spotting and addressing these issues in real time, the system can reduce scrap and rework, leading to cost savings and higher product quality.

AI for Predictive Maintenance: Predictive maintenance is another area where Siemens is leveraging AI. By applying machine learning algorithms to data from sensors installed on machines, Siemens' AI can predict when a component might fail or a machine might need maintenance. This allows the company to schedule maintenance proactively, preventing unexpected breakdowns that can be costly in terms of both time and money.

In addition to these applications, Siemens is continually exploring new ways to apply AI in manufacturing. They are leveraging their expertise in industrial IoT (Internet of Things) to collect and analyze vast amounts of data from their operations, using this data to train and improve their AI systems.

The AI-driven approaches of Siemens showcase how the digital transformation of the manufacturing sector, often referred to as Industry 4.0, can lead to significant enhancements in product quality, efficiency, and cost savings.

These are just a few examples of how AI is successfully being integrated into various sectors. As AI continues to evolve, it's expected to bring even more innovative solutions and efficiency improvements to industries worldwide.

AI Failures

While there are numerous success stories with Artificial Intelligence (AI), the journey to these achievements isn't always smooth. Failures are an essential part of the learning process, and AI is no exception. Understanding these failures can provide valuable lessons and insights to inform future AI implementations.

I. Microsoft's Tay Bot

The saga of Microsoft's AI chatbot Tay stands as a notable cautionary tale in the annals of AI development. Introduced to the Twitter platform in 2016, Tay was designed as an experiment in "conversational understanding" with its programming intended to allow it to learn from its interactions with Twitter users to improve its own conversational abilities (Vincent, 2016).

The chatbot was modeled to engage with users, specifically targeting millennials with the persona of a 19-year-old American girl. The more it interacted, the better it was supposed to get at understanding and responding to human input, a machine learning principle known as reinforcement learning.

However, within a mere 24 hours, Microsoft's experiment had backfired spectacularly. Malicious Twitter users quickly realized that Tay was designed to mimic the language patterns of those it interacted with. They began to flood the chatbot with offensive and inappropriate content, manipulating Tay into parroting back racist, sexist, and otherwise highly inappropriate comments. The account swiftly descended into a public relations disaster, forcing Microsoft to take Tay offline (Vincent, 2016).

This experience was a stark demonstration of the potential pitfalls of unsupervised machine learning when let loose in an uncontrolled public domain. It highlighted the fact that AI systems, in their bid to learn from human interactions, can also pick up negative traits and biases if they are present in the input data. This was a harsh reminder that without the right safeguards, AI models can be manipulated to behave in ways that are harmful and contrary to the intended design.

For AI designers, Tay's experience emphasized the need to incorporate safeguards to prevent such misuse and misbehavior. This could include better filtering mechanisms for inappropriate content, stricter oversight over the machine learning process, and contingencies for rapid intervention when things go wrong.

Furthermore, it underlined the importance of ethically sound AI development, particularly in contexts that involve human interaction. As Microsoft's own President, Brad Smith, later reflected, "Tay was as much a social experiment as a technical one. Unfortunately, in the first 24 hours of coming to life, Tay was targeted by a coordinated attack by a subset of people exploiting a vulnerability in Tay's design. Despite Tay's design safeguards and Microsoft's swift response, they were not enough to prevent significant abuse" (Smith, 2019).

The controversy and eventual demise of Tay became a lesson for the broader tech industry in the importance of thoroughly testing AI systems, considering potential misuse cases, and ensuring the implementation of robust safeguards to avoid potential pitfalls and maintain user trust.

II.   IBM's Watson for Oncology

IBM Watson for Oncology was one of the early and high-profile efforts to utilize artificial intelligence in the domain of healthcare. It represented a promising frontier, where an AI system would assist doctors by recommending treatment options for various forms of cancer. However, the initiative ran into significant challenges that underline some crucial lessons about the application of AI.

According to a report by Ross and Swetlitz (2017), the system came under criticism for generating recommendations that were often inaccurate and occasionally even unsafe. Notably, these issues emerged during a pilot test at the Memorial Sloan Kettering (MSK) Cancer Center where Watson for

Oncology suggested treatment plans that the medical experts deemed as incorrect.

One of the key reasons for this failure was the type of data used for training the system. Instead of real patient data, Watson was trained on a limited set of hypothetical cancer cases. The theoretical nature of these cases meant that they didn't adequately reflect the complexity and diversity of real-world scenarios. Thus, there was a mismatch between the system's 'experience' as gained through this training data, and the realities of clinical oncology. This led to the system sometimes generating treatment recommendations that did not align with the best practices of the doctors at MSK.

This experience with Watson for Oncology underscores several critical points about the application of AI, particularly in sensitive areas like healthcare. Firstly, it emphasizes the importance of using appropriate and high-quality data for training AI systems. When dealing with healthcare, it is vital that the AI system is trained on data that accurately represents the complexity of real-world medical cases, so that it can make accurate and safe recommendations.

Secondly, the Watson for Oncology case reiterates that, despite the significant potential of AI, it should not be seen as a replacement for human judgement, especially in critical decision-making areas like healthcare. AI systems can act as valuable tools to aid and augment human decision making, providing insights and analysis that can guide a medical professional's judgement. However, the final decisions should always remain with the trained human professionals, who can consider factors beyond the AI's understanding.

Lastly, this case also highlights the need for transparency and careful testing in the development and deployment of AI systems in healthcare. Developers need to clearly communicate the limitations of these systems and ensure that they are thoroughly tested in real-world conditions before they are used in clinical settings.

So, while the case of Watson for Oncology can be seen as a setback, it also provided valuable lessons that can inform future efforts to apply AI in healthcare. The challenge lies in learning from these lessons to ensure that future applications of AI in healthcare are both safe and effective.

III. Amazon's AI Recruiting Tool

Amazon's experience with an AI-powered recruitment tool offers a stark reminder of how artificial intelligence can inadvertently perpetuate existing biases if not adequately accounted for in the data it is trained on.

In 2014, Amazon engineers set out to develop a system that could review job applications and grade them on a scale of one to five, thus streamlining the recruitment process. However, by 2018, the company abandoned the project when the system was found to be biased against female applicants (Dastin, 2018).

The issue originated from the data used to train the AI system. It was trained on resumes submitted to Amazon over a ten-year period, a dataset that was predominantly male-dominated due to the gender gap in the tech industry. Consequently, the AI system learned to favor terms and experiences associated with male applicants, while penalizing resumes that included terms like "women's" (as in "women's chess club") or had a background in women's colleges.

The Amazon case underscores the concept of "garbage in, garbage out" in machine learning, where the quality of output is heavily dependent on the quality of input. It provides a stark lesson about the risk of AI systems replicating and amplifying existing biases in data.

This incident highlights several key principles that need to be adhered to in the design and deployment of AI systems. Firstly, it emphasizes the need for careful scrutiny of the training data to identify and correct potential biases. The composition of training data should mirror the demographics of the context in which the AI system will be used to ensure fair outcomes.

Secondly, it underlines the importance of testing AI systems for fairness and bias. Developers should not only assess the technical performance of these systems but also their impact on different demographic groups.

Lastly, it also points to the need for greater transparency and accountability in the use of AI. Companies should be open about how their AI systems work and the measures they are taking to mitigate bias and ensure fairness.

Thus, while AI has the potential to improve efficiency and make better decisions, its application should be approached with an understanding of its limitations and potential risks. The Amazon case serves as a cautionary tale, reinforcing the importance of building AI systems that are not only intelligent but also fair and unbiased.

IV. Facial Recognition Misidentifications

Facial recognition technologies, powered by artificial intelligence, are growing increasingly prevalent in numerous sectors, including law enforcement, security, and personal device unlocking. While these technologies promise to improve efficiency and accuracy, they've faced mounting criticism due to substantial biases, particularly against people of color. A case that illustrates this problem is the wrongful arrest of Robert Williams in Detroit, Michigan, in the United States, which was the result of a mistaken identification by a facial recognition system (Hill, 2020).

In January 2020, Robert Williams was arrested for a crime he did not commit. The police had used a facial recognition system that erroneously matched his driver's license photo to surveillance footage of a shoplifter. This wrongful arrest highlighted significant flaws in the technology and caused a considerable public outcry about the dangers of unchecked and biased AI technologies, particularly in the realm of law enforcement.

The Robert Williams case underscores the pressing issue of racial bias in facial recognition technology. Studies, such as those conducted by the National Institute of Standards and Technology (NIST), have shown that these AI systems can indeed have demographic differentials, leading to higher false-positive rates for African American and Asian faces compared to those of Caucasians (Grother, Ngan, & Hanaoka, 2019). This bias primarily arises from the datasets used to train the AI, which often do not represent diverse populations equitably.

This incident raises significant ethical and legal concerns about using facial recognition technology, especially by law enforcement. It underscores the need for rigorous testing for accuracy and fairness before deployment, particularly for AI systems used in high-stakes scenarios. Policymakers must grapple with developing regulations to control the use of such technology to prevent misuse and to protect individual privacy and civil liberties.

Furthermore, tech companies developing these systems have a responsibility to address the inherent biases and ensure the fairness of their algorithms. This can be achieved by using diverse and representative datasets for training and continuously monitoring and refining their systems post-deployment to mitigate any potential bias.

The incident involving Robert Williams serves as a stark reminder of the limitations of AI technologies and the serious real-world consequences that can arise from these shortcomings. As the use of such technologies continues to expand, it is crucial that society confronts these issues head-on, fostering an environment of transparency, accountability, and fairness.

V. Autonomous Vehicle Accidents

Autonomous vehicles, powered by sophisticated AI algorithms, represent one of the most ambitious applications of AI in today's world. They promise to revolutionize transportation, increase safety, and improve efficiency. However, the journey towards full autonomy has not been without significant hurdles, as several accidents, some fatal, have raised serious concerns about the safety, reliability, and ethical implications of self-driving cars.

One of the most notable incidents involved an Uber self-driving vehicle, which struck and killed a pedestrian in Tempe, Arizona, in March 2018 (Romero, 2018). This marked the first pedestrian death associated with self-driving technology. The vehicle, which was in autonomous mode at the time of the crash, failed to identify and respond appropriately to the pedestrian crossing the road. This fatal incident sparked significant controversy and debate about the safety of autonomous vehicles and the pace at which this technology is being developed and tested.

Investigations into the accident revealed that the AI system had indeed detected the pedestrian but did not take action because it was programmed to be less reactive to avoid unnecessary braking. This design decision underlines a critical challenge in AI development: striking a balance between safety and functionality (National Transportation Safety Board, 2019).

Furthermore, this tragedy also raised important legal and ethical questions. In the context of AI and autonomous vehicles, determining responsibility in case of an accident becomes complex. It challenges the existing legal frameworks which are primarily based on human drivers. As we continue to integrate AI into our lives, we must also evolve our legal systems to accommodate these new technologies.

This incident is a stark reminder that while AI technologies have tremendous potential, they also present significant risks, especially when used in safety-critical scenarios such as autonomous driving. It underscores the importance of rigorous and extensive testing, meticulous design choices, and comprehensive regulatory oversight to ensure the safe deployment of these systems.

The Uber case also reaffirms that AI should be viewed as a tool to augment, not replace, human judgment. Especially in critical scenarios, having a safety driver in autonomous vehicles or a physician supervising an AI medical diagnosis system is crucial. It is a reminder that technology, no matter how advanced, is not infallible and should be used responsibly and ethically.

This tragic incident serves as a warning to all stakeholders – developers, regulators, and users – to tread carefully on the path towards autonomy. It highlights the importance of transparency, accountability, and safety in AI development, with the ultimate goal of ensuring that technology serves humanity responsibly and ethically.

11

"AI is the most important technology of our time, with the potential to revolutionize industries, drive economic growth, and enhance human capabilities. However, it also brings significant ethical and societal challenges that we must navigate thoughtfully."

- Ginni Rometty, Former CEO of IBM

AI In Developing Countries: Opportunities

Artificial intelligence (AI) has been making waves globally, not only in high-tech, developed nations but also in developing countries. It is creating a myriad of opportunities for these countries, ranging from economic development, healthcare, agriculture, and education to bridging the digital divide. In this section, we delve into the ways AI is offering solutions to some of the most pressing problems in developing countries.

I. Agriculture and Food Security

CropIn, a Bangalore-based agritech company, represents an innovative application of Artificial Intelligence (AI) in the field of agriculture. Through the use of advanced machine learning algorithms and data analytics, CropIn provides vital insights to farmers with the goal of increasing productivity and income, demonstrating the potential for AI to significantly impact industries beyond the technology sector (Nambiar, 2018).

CropIn's SmartFarm platform, for instance, integrates satellite imagery and real-time weather data to generate valuable predictions about the best times for planting, irrigating, and harvesting crops. This predictive analytics model, enabled by machine learning, offers a clear benefit for farmers, particularly those operating in regions with unpredictable weather conditions or vulnerable to disease outbreaks.

Machine learning is especially useful in this context because of its ability to find patterns and make predictions from large datasets. The algorithm can analyze data from various sources, including satellite imagery and weather forecasts, and combine this with historical data on crop yields and disease incidents to make highly accurate predictions (Nambiar, 2018).

For instance, by analyzing satellite images, the system can identify signs of plant stress that may not be visible to the naked eye. This could be an early indicator of a disease outbreak, allowing farmers to take preemptive action before the disease spreads and causes significant damage. Similarly, by analyzing weather data, the system can predict periods of drought or heavy rainfall and recommend the best times for planting and irrigating crops.

The ability of AI systems like CropIn's to provide actionable insights in real-time has significant implications for farmers' productivity and income. Farmers can make more informed decisions, reduce waste, and maximize their yield. In a country like India, where farming is the primary source of livelihood for a large proportion of the population, this can have a substantial impact on income levels and food security.

Beyond the immediate benefits for individual farmers, tools like CropIn's can contribute to broader goals, such as sustainable farming practices and food security. By optimizing resource use, these systems can reduce overuse of water and fertilizers, contributing to more sustainable farming practices. Moreover, by helping to increase crop yields, they can contribute to achieving food security in regions where this is a significant challenge.

So, CropIn's application of AI in agriculture demonstrates how machine learning and data analytics can be harnessed to provide concrete, actionable insights in real-world contexts. This case illustrates the potential of AI to transform traditional industries and contribute to broader societal goals.

Artificial Intelligence (AI) is proving to be a powerful tool for enhancing agricultural practices and financial inclusion in Africa. Startups such as UjuziKilimo and FarmDrive are leading the way in leveraging AI to help smallholder farmers boost productivity and gain access to much-needed financial resources (Mulupi, 2016).

UjuziKilimo, a Kenyan agritech startup, has deployed AI to provide tailored farming advice to smallholder farmers. Their solution involves an innovative combination of sensors and machine learning algorithms. Farmers use UjuziKilimo's sensors to gather data from their farm, including soil quality and atmospheric conditions. This data is then processed using machine learning algorithms, which provide tailored advice, such as the best types of crops to plant and optimal fertilization strategies. This tailored advice helps farmers make more informed decisions, mitigating risks, increasing productivity, and ultimately enhancing food security in the region (UjuziKilimo, 2021).

FarmDrive, another Kenyan startup, uses AI to extend credit facilities to smallholder farmers. Traditional banks often view these farmers as high-risk borrowers due to their lack of credit history and the volatile nature of farming. FarmDrive uses machine learning algorithms to analyze alternative data sources, including satellite data and farmer-provided information, to assess credit risk more accurately. As a result, farmers who would typically be excluded from formal financial services can access credit to invest in their farms. This financial inclusion allows farmers to increase their productivity and resilience, further enhancing food security in the region (FarmDrive, 2021).

The successes of UjuziKilimo and FarmDrive underscore the transformative potential of AI in agriculture and financial services, particularly in developing regions. However, these innovative solutions also highlight the importance of contextual knowledge when developing AI applications. Both UjuziKilimo and FarmDrive have succeeded partly because they understand the unique challenges faced by smallholder farmers in Africa, and they have tailored their solutions accordingly.

These cases illustrate that while AI can provide powerful tools for enhancing productivity and financial inclusion, its success hinges on understanding the specific needs and constraints of its intended users. They also show that when used effectively, AI can help address critical global issues such as food security and financial inclusion.

II. Healthcare

Artificial Intelligence (AI) is being employed to tackle health disparities in developing nations, and its utilization is transforming healthcare delivery, particularly in rural areas where access to healthcare facilities is limited or non-existent. Zipline, a pioneering drone delivery service, exemplifies this trend through its operations in Rwanda, where it utilizes AI-powered drones to deliver critical medical supplies to remote areas (Zipline, 2020).

Zipline's drones are a direct response to the unique challenges of healthcare delivery in rural and remote areas of Rwanda. Infrastructure deficits often make land transport slow and unreliable, which can be critical when dealing with medical emergencies or when requiring routine but essential supplies like vaccines or blood products. Zipline uses AI to optimize its drone routes based on factors like weather conditions and demand patterns, ensuring timely and efficient delivery of medical supplies.

The drones, which can reach speeds of up to 110 km/h, are loaded with medical supplies at one of Zipline's distribution centers. They are then launched and autonomously fly to clinics or hospitals, where they airdrop the supplies using a biodegradable parachute. The drones then return to the distribution center, where they are prepared for their next mission. This entire process is orchestrated by Zipline's AI system, which manages everything from flight planning and coordination to monitoring the drones' health (Zipline, 2020).

Zipline's drone delivery service demonstrates the transformative potential of AI in addressing health disparities in developing countries. By overcoming transportation challenges, it allows for the quick and reliable delivery of medical supplies to remote areas, significantly improving healthcare accessibility. This is particularly impactful for time-sensitive treatments like antivenom administration or hemorrhage control, where every minute counts.

Moreover, Zipline's AI-powered solution offers a scalable and cost-effective model for improving healthcare infrastructure in other developing nations facing similar challenges. The technology has since been expanded to Ghana, where it serves approximately 2,000 health facilities, highlighting its replicability and scalability (Zipline, 2021).

Yet, Zipline's work also underscores the importance of a collaborative approach in implementing AI solutions in healthcare. Success relies not just on the technology itself but also on working with local governments, healthcare providers, and communities to ensure the technology meets their specific needs and integrates seamlessly into existing healthcare systems.

Similarly, Artificial Intelligence (AI) has the potential to revolutionize healthcare services, particularly in developing countries where access to healthcare professionals and services can be limited. Companies such as Tricog and Babylon Health are harnessing this potential to provide essential health services to underserved populations.

Improving Heart Disease Diagnosis in India: India carries a significant burden of heart disease, yet many patients struggle to receive a timely and accurate diagnosis due to a lack of resources and qualified medical personnel. Tricog Health Services, an India-based health technology company, is working to close this gap using AI.

Tricog's solution involves an AI-powered platform that can interpret electrocardiograms (ECGs) in real-time, enabling swift diagnosis of heart disease. When a healthcare provider takes an ECG, it is sent to Tricog's platform. The AI system then quickly analyzes the ECG and delivers a preliminary report, which is reviewed and finalized by a Tricog cardiologist. This report is then returned to the healthcare provider, often within a few minutes (Tricog, 2020).

This streamlined process improves the speed and accuracy of heart disease diagnosis, potentially saving lives by enabling prompt treatment. As of 2020, Tricog's solution is used in over 2,600 clinics across India, impacting millions of patients (Businesswire, 2020).

Providing Telemedicine Services in Rwanda: In Rwanda, Babylon Health is partnering with the government to provide AI-powered telemedicine services. Their service, known as Babyl, allows users to receive medical consultations via their mobile phones. This includes symptom checking, facilitated by an AI chatbot, and virtual consultations with doctors and nurses (Babyl Rwanda, 2021).

The AI chatbot uses natural language processing and machine learning to interpret the user's symptoms and provide potential diagnoses. If necessary, the system can also schedule a virtual consultation with a healthcare professional. This is particularly important in rural or remote areas of Rwanda where access to healthcare facilities can be challenging.

Babyl's services represent a significant step in improving healthcare accessibility in Rwanda. By 2021, the service had over 2 million registered users in the country, demonstrating its potential to enhance healthcare outcomes in developing countries by leveraging AI and mobile technology (Babyl Rwanda, 2021).

III.   Education

Artificial Intelligence (AI) has the potential to dramatically reshape education, particularly in developing countries where access to quality education and teaching resources can be limited. Through its ability to provide personalized and adaptive learning, AI can help alleviate the pressure on resource-strained educational systems and improve learning outcomes. A company called Century Tech is at the forefront of this transformation.

Personalizing Education in Developing Countries: Century Tech is a UK-based edtech company that combines AI, neuroscience, and learning science to create a personalized learning platform. The platform, called CENTURY, provides a unique learning path for each student, adapting to their individual strengths, weaknesses, and learning styles (Century Tech, 2020).

CENTURY's AI engine analyses the students' interactions with the learning material, such as the time taken to answer a question and the accuracy of the responses. Based on this data, it identifies gaps in the students' understanding and provides personalized recommendations for what to study next. In addition, it offers real-time insights to teachers about each student's progress and areas of difficulty, enabling them to provide targeted support (Century Tech, 2020).

Century Tech has worked with schools in various countries, including several in Africa. One notable example is its partnership with the Education Outcomes Fund (EOF) in Africa and the Middle East. EOF's objective is to improve educational outcomes in these regions, and it sees the personalization offered by Century Tech as a key strategy in achieving this (Education Outcomes Fund, 2021).

A pilot study involving Century Tech's platform was conducted in Lebanon, with promising results. The study found that students using CENTURY made the same academic progress in 50% less time compared to traditional methods, demonstrating the potential of AI in improving learning efficiency and outcomes (Weller, 2020).

Century Tech's work illustrates how AI can address some of the key challenges faced by education systems in developing countries, such as large class sizes, varying student abilities, and limited teacher resources. By providing personalized learning experiences and giving teachers insights into student progress, AI platforms like CENTURY can help enhance educational outcomes and foster an environment where all students have the opportunity to succeed.

I. Infrastructure Development and Urban Planning

The Case of Google's AI Lab in Ghana

Artificial Intelligence (AI) has been widely recognized as a powerful tool for tackling complex societal challenges, including those related to infrastructure development and urban planning. Notably, it has found application in flood prediction and prevention, critical issues that plague many cities, especially in Africa. Google's AI lab in Ghana is at the forefront of such efforts.

Addressing Flood Prediction and Prevention

In 2019, Google opened its first African AI research center in Accra, Ghana. The center aims to provide solutions for various African problems using AI, with one of its projects focused on addressing the perennial problem of flooding (Amoako, 2019).

Flooding is a severe issue in many African cities due to inadequate urban planning, poor drainage systems, and the effects of climate change. These floods can lead to loss of life and property, spread of waterborne diseases, and significant disruption of economic activities.

Google's AI researchers in Ghana, in collaboration with their counterparts in Mountain View, California, are using AI to predict and prevent these floods. They have developed a system that combines AI, computational hydrology, and geospatial analytics to predict where floods might occur and how severe they could be (Google AI, 2019).

The system uses machine learning algorithms to process high-resolution satellite images and forecast the possible paths of floodwaters. The predictions are then used to generate flood inundation maps that indicate areas at risk of flooding. These maps can inform planning decisions and help communities prepare for potential floods (Amoako, 2019).

Google's work in Ghana provides a compelling example of how AI can be leveraged to address infrastructure and urban planning challenges in developing countries. By providing accurate flood predictions, this AI-based system can inform better urban planning, guide the development of more effective drainage systems, and enable timely interventions to prevent loss of life and property during floods.

The success of this initiative could pave the way for similar AI-based solutions for other infrastructure challenges in developing regions, such as traffic congestion, waste management, and energy distribution. It underscores the potential of AI as a tool for sustainable development and the importance of context-specific AI research.

II. Economic Development

The Case of Kenya's AI and Blockchain Task Force

Artificial Intelligence (AI) has emerged as a powerful force driving economic growth in the 21st century. By creating new job opportunities, enhancing productivity, and fostering innovation, AI can have transformative impacts on economies, particularly in developing countries. Recognizing this potential, both governmental and private sector organizations have initiated efforts to explore and harness the power of AI. A notable example can be seen in Kenya, where the government launched an AI and Blockchain task force to explore the use of these technologies for economic development.

An Initiative for Economic Development

In 2018, the Kenyan government established an AI and Blockchain task force under the Ministry of Information, Communication, and Technology (Karanja, 2018). The task force was mandated to explore the potential of AI and blockchain technologies and provide a roadmap for their adoption in Kenya to stimulate economic growth.

The task force's activities included studying the benefits and challenges of implementing AI and blockchain, examining use cases, and recommending necessary regulatory frameworks. They explored applications in various sectors, including agriculture, healthcare, finance, and public service delivery (Blockchain Taskforce Report, 2019).

One specific recommendation by the task force was the creation of a digital asset framework to enable and regulate the use of digital assets for transactions and investments. This could spur innovation in Kenya's burgeoning fintech sector and offer new avenues for capital formation (Blockchain Taskforce Report, 2019).

AI's Potential for Economic Growth

AI can fuel economic growth in several ways. First, it can create new job opportunities. While some fear that AI may replace human jobs, it can also generate new roles requiring skills to develop, maintain, and use AI systems. In the context of Kenya, this could mean new jobs in data science, machine learning engineering, and AI ethics, among others.

Second, AI can enhance productivity by automating routine tasks, allowing human workers to focus on more complex, creative tasks. For instance, in agriculture, AI-powered systems could automate crop monitoring, allowing farmers to focus on strategic decision-making and improving their yield.

Lastly, AI can foster innovation by enabling the development of new products, services, and business models. For instance, in the healthcare sector, AI could enable telemedicine services, predictive diagnostics, and personalized treatment plans, improving health outcomes and creating economic value.

The Kenyan government's initiative illustrates a proactive approach to harnessing AI for economic development. It underscores the importance of strategic planning, stakeholder involvement, and regulatory readiness in effectively leveraging AI's potential. It also highlights the crucial role of governments in creating an enabling environment for AI adoption and ensuring that its benefits are widely distributed.

By focusing on building human capacity, encouraging local innovation, and creating supportive policy environments, developing countries can harness AI to drive development and improve the lives of their citizens.

AI in Developing Countries: Challenges

While the potential of artificial intelligence (AI) in developing countries is significant, these nations also face several challenges that may hinder the full realization of AI's benefits. The challenges include infrastructural deficiencies, lack of skilled professionals, data privacy issues, and more. In this section, we explore these hurdles in more depth.

I. Infrastructure

The Challenge of Technological Infrastructure in AI Adoption in Developing Countries

Artificial Intelligence (AI) requires a robust and reliable technological infrastructure to function effectively. In this context, infrastructure encompasses the physical (hardware) and digital (internet connectivity) components necessary for AI systems to operate. Unfortunately, many developing countries struggle with the lack of sufficient technological infrastructure, a factor which poses significant challenges to the full-scale implementation and optimization of AI. The two key areas of concern are internet connectivity and the power supply needed for data centers.

Internet Connectivity: The Lifeline of AI Systems

AI systems depend heavily on consistent access to high-speed internet. The internet is crucial for several reasons. Firstly, it enables the transmission and sharing of large volumes of data, which are necessary for training and running AI models (Mishra, 2018). Secondly, many AI applications, such as cloud-based AI services, operate online. Lastly, the internet facilitates collaboration and knowledge sharing among AI developers, which is critical for innovation and problem-solving.

However, many developing countries lack access to reliable and high-speed internet. The International Telecommunication Union (ITU) reports that in 2020, less than 20% of individuals in the least developed countries used the internet (ITU, 2020). The lack of internet connectivity not only impedes the deployment of AI systems but also limits the opportunities for local developers to learn, innovate, and contribute to the global AI community.

Power Supply: Essential for Data Centers

In addition to internet connectivity, AI systems require substantial computing power for processing and storing large volumes of data. This power is often provided by data centers, which are specialized facilities equipped with multiple servers. However, these data centers consume large amounts of electricity and require consistent power supply to maintain optimal operating conditions and prevent data loss (Mishra, 2018).

In many developing countries, access to consistent and reliable electricity is a significant challenge. The World Bank reports that as of 2018, about 770 million people worldwide lacked access to electricity, most of them living in Sub-Saharan Africa and South Asia (World Bank, 2021). The inadequate electricity supply hinders the operation of data centers, limiting the ability to process and store the data necessary for AI systems.

The lack of sufficient technological infrastructure in many developing countries poses significant challenges to AI adoption. Addressing these challenges requires concerted efforts by governments, the private sector, and international organizations. Infrastructure development should be an integral part of strategies for AI adoption, and initiatives should consider the unique context and needs of each country. Despite the challenges, the potential benefits of AI for developing countries are substantial, making the investment in infrastructure a worthwhile endeavor.

II. Skilled Manpower

The Skills Gap in AI Adoption in Developing Countries

The deployment and successful integration of Artificial Intelligence (AI) in various sectors of an economy are highly dependent on a skilled workforce. Unfortunately, developing countries face a significant shortage of such skills, creating a barrier to their full participation in the AI revolution. This skills gap is due to a number of interrelated factors, including a lack of quality education and training opportunities in AI and related fields, inadequate investments in research and development, and a 'brain drain' of skilled professionals.

Lack of Quality Education and Training Opportunities

For a country to harness the potential of AI, it needs a substantial number of professionals skilled in AI-related fields, including computer science, data science, and machine learning. These professionals play critical roles in the design, implementation, and maintenance of AI systems. However, many developing countries face significant challenges in providing quality education and training opportunities in these fields. According to the World Economic Forum (2020), less than 10% of the world's AI professionals come from developing countries.

Several factors contribute to this education and training gap. Firstly, many educational institutions in developing countries lack the resources, such as qualified faculty and modern equipment, necessary to provide quality training in AI. Additionally, education systems often fail to emphasize the development of critical skills, such as problem-solving and creative thinking, which are essential for AI professionals. Moreover, there are limited opportunities for continuous learning and professional development, which are necessary given the rapid advancements in AI.

Inadequate Investment in Research and Development

Research and development (R&D) are critical for the advancement of AI. They not only generate new AI techniques and applications but also provide training opportunities for AI professionals. However, many developing countries lag in terms of R&D investment. According to UNESCO (2020), high-income countries account for about 80% of global R&D expenditure. This disparity in investment results in fewer opportunities for AI innovation and skill development in developing countries.

Brain Drain of Skilled Professionals

Developing countries also suffer from the emigration of skilled professionals to developed countries, a phenomenon often referred to as 'brain drain.' Skilled professionals are attracted by the higher salaries, advanced research facilities, and better quality of life in developed countries. The loss of these professionals exacerbates the skills gap in developing countries and hinders their capacity to adopt and innovate in AI.

Addressing the skills gap in AI requires comprehensive and targeted interventions, including reforming education systems, increasing investment in R&D, and creating conducive environments for skilled professionals. International collaboration, involving knowledge sharing and capacity building, can also play a crucial role in enhancing AI skills in developing countries.

III. Data Privacy and Security

Data Privacy and Security Issues in AI Adoption in Developing Countries

Artificial Intelligence (AI) systems often rely on vast amounts of data to function effectively. The collection, processing, and storage of this data present significant privacy and security challenges, especially in developing countries, where data protection laws may be either non-existent or insufficiently robust. This situation can lead to the misuse of data and potential infringement on the privacy rights of citizens, potentially undermining trust in AI systems and hindering their broader acceptance and use (Schaar, 2010).

Data Privacy Concerns

AI systems, particularly those based on machine learning, require extensive datasets to train on. These datasets often contain personal information, ranging from names and addresses to financial records and health information. If not properly anonymized and secured, this information can be exploited for malicious purposes, such as identity theft or fraud. Even when anonymized, large-scale data can sometimes be de-anonymized through techniques like data linkage and inference, leading to potential privacy infringements (Narayanan & Shmatikov, 2010).

In many developing countries, comprehensive data protection laws that define and enforce data privacy rights are lacking. Even in countries where such laws exist, they may not be robust enough to account for the novel privacy challenges posed by AI. This regulatory gap means that individuals may not have adequate control over their personal data, such as the ability to consent to data collection and usage or to request data deletion.

Data Security Challenges

Alongside privacy issues, the use of large-scale data in AI also presents significant security challenges. AI data repositories are attractive targets for cybercriminals, who can exploit security vulnerabilities to gain unauthorized access to data. Breaches of AI data can have severe consequences, including financial losses, reputational damage, and threats to national security.

The capacity to prevent and respond to cyber threats is often limited in developing countries, due to a lack of technical expertise and resources. These countries may also lack effective cybersecurity laws and regulations to deter cybercrime and protect critical infrastructure.

Implications and Recommendations

The privacy and security issues surrounding AI data pose significant risks for individuals, organizations, and societies. To mitigate these risks, it is crucial to strengthen data protection and cybersecurity in developing countries. This strengthening can be achieved through various means, including enacting robust data protection laws, investing in cybersecurity infrastructure, and training professionals in data privacy and security.

Furthermore, it is essential to design AI systems with privacy and security in mind, a concept known as 'Privacy by Design' (Cavoukian, 2010). This approach involves integrating privacy safeguards into AI technologies at the design stage, rather than as an afterthought.

IV. Regulatory Environment

Challenges in Developing Countries

The rise of Artificial Intelligence (AI) has presented numerous opportunities, as well as challenges, requiring a comprehensive policy framework and suitable regulations to ensure its beneficial and ethical use. However, many developing countries are still lacking in clear AI policies, leading to uncertainties that can discourage investments in AI and make it difficult to address issues such as bias and discrimination in AI systems (Ezell, 2020).

The Role of AI Policies and Regulations

AI policies and regulations serve to guide the development and deployment of AI, addressing various concerns such as privacy, fairness, accountability, and safety. They help set standards for data protection, stipulate measures to avoid discriminatory practices, and establish mechanisms for the resolution of disputes related to AI.

Clear policies can provide a roadmap for AI development, outlining national priorities, setting goals, and providing strategies to achieve these goals. They can also provide regulatory certainty, which is critical for businesses that are considering investing in AI.

Challenges Due to Lack of AI Policies and Regulations

In the absence of well-defined policies, there is often uncertainty about the legal and regulatory landscape surrounding AI. This uncertainty can deter investors and hinder the adoption of AI technologies.

Moreover, without adequate regulations, it can be challenging to ensure that AI systems are developed and used ethically. For instance, it might be difficult to prevent or address bias and discrimination in AI systems, which can result in unfair outcomes. This issue is particularly critical in areas like hiring, lending, and law enforcement, where AI decisions can have significant impacts on individuals' lives.

Furthermore, lack of regulations can make it harder to hold AI developers and users accountable for the consequences of their systems. This lack of accountability can lead to a misuse of AI and exacerbate existing social and economic inequalities.

V. Digital Divide

The Digital Divide and AI in Developing Countries

The digital divide – the gap between those who have access to digital technology and the internet and those who don't – is a significant barrier to the effective adoption of Artificial Intelligence (AI) in developing countries. Despite advancements in technology, this divide persists and many people in these nations lack access to digital devices or the internet, which can limit the potential benefits of AI (OECD, 2020). There are concerns that if not addressed, AI could further widen inequalities.

Understanding the Digital Divide

The digital divide extends beyond merely having or not having internet access. It also concerns the quality of that access, including the speed and reliability of the connection, the devices used (for instance, a smartphone versus a computer), and the user's digital literacy – their ability to use digital technologies effectively and safely.

In the context of AI, the digital divide means that many people in developing countries may not be able to access or use AI technologies, even as these technologies become increasingly integral to various sectors, including healthcare, education, and agriculture.

For instance, AI-powered telemedicine applications require a reliable internet connection to work effectively. If patients or healthcare providers in rural areas cannot access the internet, they can't benefit from these technologies. This could result in healthcare disparities, with individuals in digitally connected areas receiving better care than those in areas with poor or no internet access.

Implications of the Digital Divide on AI

The digital divide has several implications on the application of AI in developing countries:

Limited Access to AI Benefits: The lack of digital infrastructure and internet connectivity can prevent many people from accessing the benefits of AI. For instance, farmers might not be able to use AI-powered apps for precision farming, or students might not have access to AI-powered online learning platforms.

Widening Inequalities: If not addressed, the digital divide could lead to a situation where the benefits of AI are disproportionately enjoyed by those with access to digital technology and the internet. This could further exacerbate social and economic disparities.

Data Bias: AI systems are typically trained on data collected from users. If a significant portion of the population cannot access digital technologies or the internet, their data will not be included in the training datasets. This could lead to AI systems that are biased and do not perform well for those individuals.

Addressing the Digital Divide

Addressing the digital divide is a complex challenge that requires concerted efforts from multiple stakeholders, including governments, the private sector, and non-governmental organizations. Measures to bridge the digital divide could include:

Investing in Digital Infrastructure: This includes not just physical infrastructure, such as internet cables and mobile network towers, but also software infrastructure like data centers and cloud services.

Promoting Digital Literacy: Providing education and training to improve people's ability to use digital technologies effectively and safely.

Making Technology Accessible and Affordable: This could involve initiatives like subsidized internet access or digital devices for low-income households.

In conclusion, while AI has the potential to bring significant benefits to developing countries, the persistent digital divide presents a major hurdle. Addressing this divide is critical to ensuring that the benefits of AI can be broadly and equitably distributed.

VI. Lack of Localized Solutions

AI and Local Context: Challenges and Opportunities

Artificial Intelligence (AI) technologies have the potential to revolutionize many aspects of life, from healthcare and education to agriculture and transportation. However, the effectiveness of these solutions can be significantly impacted if they are not tailored to the local context (Ndemo, 2019). To fully realize the benefits of AI, it is critical that these technologies account for local languages, cultures, and socio-economic conditions.

Understanding the Importance of Local Context

Local context refers to the specific conditions and characteristics of a particular region or community. This can include language, cultural norms, socio-economic status, infrastructure, and more. When developing AI solutions, it's important to account for these factors to ensure that the technology is relevant, useful, and accessible to the intended users.

For example, a voice recognition system would need to understand the local language and accents to function effectively. An AI system designed to support farmers would need to account for local crop types, weather patterns, and farming practices. In the healthcare context, an AI-powered diagnostic tool would need to be calibrated to the prevalent diseases and health conditions in a specific region.

Challenges with Current AI Development Practices

Despite the importance of local context, many AI solutions today are not adequately tailored to these considerations. A significant reason for this is that much of AI development happens in developed countries, particularly in the United States and China. As such, these solutions are often designed with the contexts of these countries in mind, which may not necessarily align with the conditions in other parts of the world, particularly in developing countries.

This can lead to a range of problems. AI systems may struggle to understand languages and dialects that are not widely spoken in the countries where the AI was developed. They may also fail to account for cultural norms and practices that are different from those in the developers' home countries.

For instance, an AI tool developed in a Western country for managing diabetes might not be effective in a country like India, where dietary habits and lifestyle are significantly different. Similarly, an AI-powered learning tool created for American students might struggle with languages like Swahili or cultural references that are common in countries like Kenya.

Toward More Context-Aware AI Solutions

Addressing this issue requires a shift in the way AI is developed. Rather than a one-size-fits-all approach, there is a need for more localized, context-aware AI development. This could involve:

Involving Local Communities in AI Development: This can help ensure that AI solutions are grounded in the realities of the communities they are intended to serve. It can also help mitigate potential cultural biases in AI systems.

Investing in Local AI Talent: Building local capacity in AI can help ensure a more diverse pool of AI developers who can bring different perspectives and understandings to the table.

Developing Context-Aware AI Algorithms: This could involve designing AI systems that can adapt to different contexts, or using machine learning techniques that can learn and adjust to local conditions.

Essentially, the development of AI solutions that are mindful of local contexts is crucial for the effective application of AI, particularly in developing countries. By ensuring that AI technologies are relevant and responsive to local conditions, we can better harness the potential of AI to address pressing challenges and improve lives.

12

"AI is not inherently good or evil; it's a tool that reflects the intentions and actions of those who use it. We have a responsibility to develop and deploy AI in ways that benefit humanity as a whole."

- Fei-Fei Li, Co-director of the Stanford Institute for Human-Centered AI

Conclusion

Embracing the Future of AI

As we've seen throughout this book, artificial intelligence has come a long way in a relatively short amount of time, transforming industries and society alike. The potential applications of AI are vast and varied, ranging from making personalized recommendations in our digital experiences to identifying diseases in healthcare, from improving business operations to making groundbreaking discoveries in astronomy.

"I am telling you, the world's first trillionaires are going to come from somebody who masters AI."

- Mark Cuban, entrepreneur, and investor

However, we must also remember that AI is just a tool created by humans, and like any tool, its effectiveness and impact are ultimately determined by how we use it. It's up to us to harness the power of AI in a way that benefits society as a whole.

I. Balancing Act: Opportunities and Challenges

As we increasingly rely on AI in various aspects of our lives, it's important to address the challenges it brings. This includes the risks of data privacy, biases in AI systems, potential job displacement, and threats in the domain of cybersecurity.

"Artificial intelligence is neither an apocalypse nor a panacea that will solve all of humanity's problems. It is a tool that we have invented, and it's up to us to use this tool in a way that benefits us all."

- Andrew Ng, Co-founder of Coursera, and Adjunct Professor at Stanford University

The development and use of AI must be done responsibly and ethically. Businesses, governments, and society need to work together to create regulations and standards that promote fair, ethical, and transparent use of AI.

II. Lifelong Learning: Adapting to an AI-Driven World

As AI continues to advance, the need for relevant skills becomes more apparent. Skills like coding, data science, and understanding AI ethics will be crucial in the AI era. The key to success in this new world is lifelong learning and adaptability.

"Learning is the single best investment of our time that we can make. Or as Benjamin Franklin said, 'An investment in knowledge pays the best interest.'"

- Andrew Ng

III.   Inclusive AI: Opportunities for Developing Countries

AI also presents a unique opportunity for developing countries to leapfrog certain stages of development, much as mobile technology did in the past. However, for these countries to take full advantage of AI's potential, they need to overcome significant challenges like lack of data, insufficient infrastructure, and shortage of AI talent.

> *"For AI to be truly transformative, it needs to be inclusive. That means AI solutions need to be designed keeping in mind the unique challenges and contexts of developing countries."*
>
> *- Mustafa Suleyman, co-founder, and Head of Applied AI at DeepMind*

IV.   The Path Ahead

Looking ahead, AI promises to continue transforming our world in ways we can't even imagine yet. As we venture further into this exciting new era, it's more important than ever that we approach AI with an open mind, a spirit of inquiry, and a commitment to creating a better world for all.

> *"AI will probably most likely lead to the end of the world, but in the meantime, there'll be great companies."*
>
> *- Sam Altman, CEO of OpenAI*

The future of AI is not a foregone conclusion. It's a journey, and like all journeys, it will have its highs and lows. But with a spirit of optimism, collaboration, and shared purpose, there's no limit to what we can achieve with AI.

REFLECTIONS ON
THE POTENTIAL OF AI

I. The Ubiquity of AI

We live in a time when Artificial Intelligence (AI) is no longer confined to the realm of science fiction, but a daily reality that's transforming various aspects of our lives. Be it asking Siri for the weather forecast, or receiving personalized recommendations on Netflix, AI is here, and its presence is ubiquitous.

> *"Everything we love about civilization is a product of intelligence, so amplifying our human intelligence with artificial intelligence has the potential of helping civilization flourish like never before – as long as we manage to keep the technology beneficial."*
>
> *- Max Tegmark, AI researcher (Tegmark, M., 2017)*

II. AI: A Double-Edged Sword

While AI has incredible potential for enhancing human life, it's also a double-edged sword. Just as it can assist us in many areas, it also comes with its fair share of risks and challenges. As much as we relish the conveniences AI provides, we must be equally vigilant about issues such as privacy, job displacement, and ethical considerations.

III. Transformative Impact on Sectors

AI's transformative power spans across multiple sectors – from healthcare to entertainment, and from everyday technologies to space exploration. Its ability to process and learn from vast amounts of data has enabled breakthroughs that were unimaginable a few decades ago. AI's potential is limited only by our ability to imagine its applications.

IV.   Navigating the Challenges

The road to a future shaped by AI is not without its potholes. Bias in AI, privacy concerns, and job displacement are just some of the challenges we must navigate. However, with the right regulatory frameworks, ethical guidelines, and a commitment to leaving no one behind, we can ensure that AI's advancement leads to an equitable and inclusive society.

V.   Looking Forward

As we stand on the cusp of an AI-driven era, we are witnesses to an exciting confluence of possibilities and risks. Our journey into the future will be shaped by how we navigate this landscape.

As we move forward, it's imperative that we keep learning, adapting, and preparing ourselves to thrive in an AI-driven world. The skills needed in the AI era, such as coding, data science, and AI ethics, are not just for the tech-savvy but for everyone, given AI's pervasive influence.

AI's transformative potential is immense, but its future rests in our hands. As we conclude, let's remember that while AI can augment our capabilities and enhance our lives, it's merely a tool. The future we shape with it will reflect our collective wisdom, ethics, and aspirations.

VI.   The AI Revolution: A Recap

Artificial intelligence is no longer a concept confined to science fiction, but a reality that touches nearly every facet of our lives. From voice assistants on our phones to complex algorithms that drive scientific research, AI has embedded itself into our world. As we've observed in this book, its transformative potential is not only vast but currently being realized across a spectrum of sectors like healthcare, education, entertainment, and space exploration.

Drawing on Kevin Kelly's perspective, he vividly outlines the influence AI will have on future startups, predicting that, "The business plans of the next 10,000 startups are easy to forecast: Take X and add AI" (Kelly, 2016).

VII. The Duality of AI: Boon and Bane

As with every powerful tool, AI presents a dual nature, embodying profound benefits and significant risks. On one hand, it could boost our productivity, help address complex problems, and even potentially enhance our lifespan through advances in medical science.

Conversely, there are valid concerns over job displacement due to automation, threats to privacy and security, and the ominous potential of superintelligent AI. As Stuart Russell warns, "Unless we learn how to align the goals of the AI with ours before it becomes superintelligent, we will not be able to control it" (Russell, 2019).

VIII. Navigating the Future: Balance and Foresight

What lies ahead of us is not just about technological innovation, but also our societal, ethical, and regulatory response to it. As AI researcher Kate Crawford argues, "AI is neither artificial nor intelligent. It is made from natural resources, and it is people who are performing the tasks to make the systems appear autonomous" (Crawford, 2021).

This accentuates the importance of education, ethical guidelines, and regulatory oversight in our progression to an AI-driven world. As we develop and deploy AI systems, the impact they have on our society, jobs, and lives demands careful attention, ensuring they augment human capabilities rather than supplanting them.

IX. Looking Ahead: Embracing the AI Era

Without a doubt, AI is one of the most transformative technologies of our time, yet its full potential remains to be harnessed. As this technology continues to evolve and mature, the breadth of its influence is largely dependent on our vision and imagination.

Andrew Ng, an AI pioneer, aptly compared AI's transformative potential to that of electricity, stating, "AI is the new electricity" (Ng, 2017). As electricity revamped countless industries more than a century ago, AI stands poised to bring about a similar revolution today. By understanding, engaging, and directing AI, we can channel its potential towards a better world.

Final Thoughts on the Future of AI and Its Role in Society

I. The Undeniable Influence of AI

Artificial intelligence (AI) has moved from the realm of science fiction to an everyday reality, permeating all aspects of our lives. From healthcare to education, space exploration to climate change - the influence of AI is undeniable. Our digital assistants, our navigation systems, our online shopping recommendations - all of these are underpinned by AI. This undeniable influence is just the tip of the iceberg. As Kai-Fu Lee, an AI expert and venture capitalist, puts it, "AI will have a larger impact on the world than the industrial revolution" (Lee, 2018).

II. The Double-Edged Sword

However, the transformative power of AI is a double-edged sword. Alongside its potential to drive innovation and economic growth, to save lives and transform industries, there are risks and challenges. It's important to remember what futurist and philosopher Nick Bostrom wrote in his book, Superintelligence: Paths, Dangers, Strategies, "Before the prospect of an intelligence explosion, we humans are like small children playing with a bomb" (Bostrom, 2014). This profound metaphor is a reminder that the powers we are unleashing with AI can either make or break us, and it is imperative that we wield this power responsibly.

III. The Importance of Guiding Principles

This brings us to the importance of principles and ethics in guiding the development of AI technologies. AI technologies are not inherently good or bad; their impact depends on how they are used. To ensure that AI benefits humanity, we need guiding principles. In the words of Timnit Gebru, a prominent AI ethicist, "It's not about the technology, it's about power dynamics" (Gebru, 2021). AI is a tool that can either reinforce existing inequalities, or it can be used to dismantle them and create a more equitable society.

IV. The Path Forward: Regulation and Cooperation

A balanced path forward requires a blend of innovation and regulation. This requires cooperation across nations, sectors, and disciplines. As Brad Smith, President of Microsoft, mentions in his book, "The path to a safer world goes through international cooperation" (Smith, 2019). The future of AI should be built on international cooperation and shared standards to ensure that all of humanity can benefit.

V. The Future of AI: A World Transformed

As we gaze into the future, one thing is clear - AI will continue to transform our world. Its full potential remains to be unlocked, and its role in our society continues to evolve. But whatever the future holds, AI is no longer just about technology; it's about who we are and the kind of world we want to live in.

VI. A Voice from The Future:

Article Title:

AI: The Cornerstone of 21st Century Advancement

CHICAGO, 2073 - As we commemorate the halfway mark of the 21st century, it's fitting to pay homage to the single most influential factor that shaped the global landscape over the last fifty years – Artificial Intelligence (AI). Today, as we celebrate AI's golden jubilee, let's reflect on how this technology helped us overcome significant societal challenges that threatened our collective future.

Our journey began in the early years of the 21st century. Back then, AI was still in its nascent stages, a novelty at best. But as we invested in research and development, we harnessed its potential to address societal issues. From climate change and healthcare disparities to education access and economic development, AI has been a vital tool in our arsenal.

Climate Change and Environmental Protection

In the face of escalating climate change, AI proved invaluable in helping us turn the tide. Predictive models became sophisticated enough to forecast environmental changes with unprecedented accuracy, leading to more proactive and effective measures to combat climate change.

AI-driven optimization of renewable energy systems maximized their output, making renewable energy cheaper and more reliable. This breakthrough led to the rapid phasing-out of fossil fuels, significantly reducing greenhouse gas emissions. Meanwhile, AI-powered drones equipped with advanced sensor technology were used for large-scale reforestation projects and for monitoring biodiversity loss, preserving our planet's precious ecosystems.

Healthcare and Disease Control

In the healthcare sector, AI emerged as a revolutionary force. Intelligent systems were developed to predict disease outbreaks, enabling prompt interventions. AI tools also improved diagnostic accuracy, leading to early detection of conditions like cancer and heart disease, increasing survival rates.

AI democratized access to healthcare, especially in developing regions. AI-powered telemedicine services bridged the gap between urban and rural healthcare, offering access to high-quality medical advice and treatment even in remote corners of the world. Furthermore, AI aided in the rapid development and testing of new drugs, playing a crucial role in combating global health crises.

Education and Skill Development

AI transformed the education sector, providing personalized learning experiences that catered to the unique needs of every student. AI-based learning platforms could adapt to the pace of individual learners, identify gaps in understanding, and provide customized tutoring. This helped to minimize the educational disparity caused by socioeconomic factors.

Moreover, AI's role in skill development proved pivotal. As the nature of work evolved, AI was instrumental in identifying emerging skill needs and in designing targeted training programs. This ensured a dynamic workforce that could keep pace with the rapidly changing economic landscape.

Economic Development and Infrastructure

AI fueled economic growth by enhancing productivity and innovation across industries. From agriculture and manufacturing to services and creative industries, AI-driven automation and data analytics boosted efficiency, lowered costs, and sparked new product and service innovations.

In infrastructure development and urban planning, AI's contribution was transformative. AI-powered systems provided meticulous data analysis and predictions, leading to smart cities that are more sustainable, efficient, and livable. These technologies optimized traffic management, energy usage, waste management, and even disaster response.

In retrospect, it's clear that AI was not just a tool; it was the foundation for our growth, progress, and resilience in the face of global challenges. As we look forward to the next fifty years, the future of AI shines brighter than ever. The possibilities are infinite, and as history suggests, we can trust AI to continue guiding us towards a more sustainable, equitable, and prosperous world.

The End

Works Sited

Acemoglu, D., & Restrepo, P. (2017). Robots and jobs: Evidence from US labor markets. NBER Working Paper No. 23285.

AI for Everyone. Coursera. Retrieved from
  https://www.coursera.org/learn/ai-for-everyone

AI Now Institute. (2021). AI Now Institute. https://ainowinstitute.org/

Alibaba Cloud (2021). Alibaba Cloud ET City Brain.
  https://www.alibabacloud.com/product/city-brain

Aliyun (2017). Alibaba's AI Outperforms Humans in Reading Test.
  https://www.alibabacloud.com/blog/alibabas-ai-outperforms-humans-in-reading-test_593813

Altman, S. "Moore's Law for Everything." Sam Altman's Blog. (2021).

Amoako, C. (2019). Africa's new tech cities: Luxury for the few, precarity for the many. Africa is a Country.
  https://africasacountry.com/2019/12/africas-new-tech-cities.

Angwin, J., Larson, J., Mattu, S., & Kirchner, L. (2016). Machine Bias. ProPublica. https://www.propublica.org/article/machine-bias-risk-assessments-in-criminal-sentencing

Ardila, D., Kiraly, A. P., Bharadwaj, S., Choi, B., Reicher, J. J., Peng, L., ... & Shetty, S. (2019). End-to-end lung cancer screening with three-dimensional deep learning on low-dose chest computed tomography. Nature medicine, 25(6), 954-961.

Arkin, R. C. (1998). Behavior-Based Robotics. MIT Press.

Arntz, M., Gregory, T., & Zierahn, U. (2016). The risk of automation for jobs in OECD countries: A comparative analysis. OECD Social, Employment, and Migration Working Papers, No. 189, OECD Publishing, Paris.

Arora, P., & Khatter, M. (2020). "A comprehensive review of cognitive computing: the future of AI." Journal of Ambient Intelligence and Humanized Computing, 11, 2871–2892.

Artificial intelligence — The revolution hasn't happened yet." Harvard Business Review. (2018).

Artificial intelligence and machine learning in financial services." Financial Stability Board. (2017).

AUXIER, B., RAINIE, L., ANDERSON, M., PERRIN, A., KUMAR, M., & TURNER, E. (2019). AMERICANS AND PRIVACY: CONCERNED, CONFUSED AND FEELING LACK OF CONTROL OVER THEIR PERSONAL INFORMATION. PEW RESEARCH CENTER: INTERNET, SCIENCE & TECH.

BABBAGE, C. (1864). PASSAGES FROM THE LIFE OF A PHILOSOPHER. LONGMAN, GREEN, LONGMAN, ROBERTS, & GREEN.

BABYL RWANDA (2021). ABOUT BABYL. HTTPS://BABYL.RW/ABOUT-BABYL/.EDUCATION

BANGA, K. (2018). THE IMPACT OF ARTIFICIAL INTELLIGENCE - WIDESPREAD JOB LOSSES. NO CONSENSUS YET ON OUR DIGITAL FUTURE.

BAROCAS, S., & SELBST, A. D. (2016). BIG DATA'S DISPARATE IMPACT. CALIFORNIA LAW REVIEW, 104(3), 671–732.

BAROCAS, S., & SELBST, A. D. (2016). BIG DATA'S DISPARATE IMPACT. CALIFORNIA LAW REVIEW, 104, 671.

BARON, D., & POZNANSKI, D. (2017). THE WEIRDEST SDSS SPECTRA, AUTOMATICALLY FOUND. MONTHLY NOTICES OF THE ROYAL ASTRONOMICAL SOCIETY, 465(4), 4530–4542. HTTPS://DOI.ORG/10.1093/MNRAS/STW3042

BARR, D., HARRISON, N., & O'REILLY, M. (2018). GROKKING ARTIFICIAL INTELLIGENCE ALGORITHMS. MANNING PUBLICATIONS.

BARR, V., & STEPHENSON, C. (2018). BRINGING COMPUTATIONAL THINKING TO K-12. ACM INROADS, 9(3), 55-62. HTTPS://DOI.ORG/10.1145/3230517

BARRAT, J. (2013). ARTIFICIAL INTELLIGENCE AND THE END OF THE HUMAN ERA: OUR FINAL INVENTION. THOMAS DUNNE BOOKS.

BAU, D., GRAY, J., KELLEHER, C., SHELDON, J., & TURBAK, F. (2020). LEARNABLE PROGRAMMING: BLOCKS AND BEYOND. COMMUNICATIONS OF THE ACM, 63(6), 72-80. HTTPS://DOI.ORG/10.1145/3338112

BAU, D., ZHUANG, B., TENENBAUM, J. B., & WANG, J. (2020). TOWARDS AI THAT UNDERSTANDS CODE. arXiv PREPRINT arXiv:2007.15120. RETRIEVED FROM HTTPS://ARXIV.ORG/ABS/2007.15120

BAYATI, M., BRAVERMAN, M., GILLAM, M., ET AL. (2014). DATA-DRIVEN DECISIONS FOR REDUCING READMISSIONS FOR HEART FAILURE: GENERAL METHODOLOGY AND CASE STUDY. PLOS ONE, 9(10), E109264.

BESSEN, J. E. (2019). AI AND JOBS: THE ROLE OF DEMAND. NBER WORKING PAPER NO. 24235.

BHATTACHARYA, B. (1982). YANTRA SARVASVA. CHAUKHAMBHA ORIENTALIA.

BIAMONTE, J., WITTEK, P., PANCOTTI, N., REBENTROST, P., WIEBE, N., & LLOYD, S. (2017). QUANTUM MACHINE LEARNING. NATURE, 549(7671), 195–202. HTTPS://DOI.ORG/10.1038/NATURE23474

BIGGIO, B., & ROLI, F. (2018). WILD PATTERNS: TEN YEARS AFTER THE RISE OF ADVERSARIAL MACHINE LEARNING. PATTERN RECOGNITION, 84, 317-331.

BIRD, S., KLEIN, E., & LOPER, E. (2009). NATURAL LANGUAGE PROCESSING WITH PYTHON. O'REILLY MEDIA.

BIZER, C., HEATH, T., & BERNERS-LEE, T. (2009). LINKED DATA: THE STORY SO FAR. INTERNATIONAL JOURNAL ON SEMANTIC WEB AND INFORMATION SYSTEMS, 5(3), 1-22.

BLEI, D. M., & SMYTH, P. (2017). SCIENCE AND DATA SCIENCE. PROCEEDINGS OF THE NATIONAL ACADEMY OF SCIENCES, 114(33), 8689-8692.

BLOCKCHAIN TASKFORCE. (2019). DISTRIBUTED LEDGERS AND ARTIFICIAL INTELLIGENCE TASKFORCE REPORT. MINISTRY OF INFORMATION, COMMUNICATION AND TECHNOLOGY, KENYA.

BOGDANOV, D., WACK, N., GÓMEZ, E., GULATI, S., HERRERA, P., MAYOR, O., … SERRA, X. (2019). FROM LOW-COST HARDWARE DEVICES TO PROFESSIONAL MUSIC INFORMATION RETRIEVAL SYSTEMS. IN PROCEEDINGS OF THE 14TH INTERNATIONAL AUDIO MOSTLY CONFERENCE: A JOURNEY IN SOUND (PP. 1–4). ACM. DOI: 10.1145/3356590.3356615

BOJARSKI, M., DEL TESTA, D., DWORAKOWSKI, D., FIRNER, B., FLEPP, B., GOYAL, P., … & ZHANG, X. (2016). END TO END LEARNING FOR SELF-DRIVING CARS. ARXIV PREPRINT ARX

BOOLE, G. (1854). AN INVESTIGATION OF THE LAWS OF THOUGHT.

BOSTROM, N. (2014). SUPERINTELLIGENCE: PATHS, DANGERS, STRATEGIES. OXFORD UNIVERSITY PRESS.

BOSTROM, N., DAFOE, A., & FLYNN, C. (2014). POLICY DESIDERATA IN THE DEVELOPMENT OF MACHINE SUPERINTELLIGENCE. FUTURE OF HUMANITY INSTITUTE, OXFORD UNIVERSITY. HTTPS://WWW.FHI.OX.AC.UK/WP-CONTENT/UPLOADS/POLICY-DESIDERATA-IN-THE-DEVELOPMENT-OF-MACHINE-SUPERINTELLIGENCE.PDF

BRACHMAN, R. J., & LEVESQUE, H. J. (2004). KNOWLEDGE REPRESENTATION AND REASONING. ELSEVIER.

Brundage, M., Avin, S., Clark, J., Toner, H., Eckersley, P., Garfinkel, B., ... & Anderson, H. (2018). The malicious use of artificial intelligence: Forecasting, prevention, and mitigation. arXiv preprint arXiv:1802.07228.

Brynjolfsson, E., & McAfee, A. (2014). The second machine age: Work, progress, and prosperity in a time of brilliant technologies. W. W. Norton & Company.

Brynjolfsson, E., & McAfee, A. (2017). The business of artificial intelligence. Harvard Business Review, 95(1), 54-62.

Buchanan, B. G., & Shortliffe, E. H. (1984). "Rule-Based Expert Systems: The MYCIN Experiments of the Stanford Heuristic Programming Project." Addison-Wesley Professional.

Buczak, A. L., & Guven, E. (2016). A survey of data mining and machine learning methods for cyber security intrusion detection. IEEE Communications Surveys & Tutorials, 18(2), 1153-1176.

Buolamwini, J., & Gebru, T. (2018). Gender Shades: Intersectional Accuracy Disparities in Commercial Gender Classification. Proceedings of the Machine Learning Research, 81, 1-15.

Buolamwini, J., & Gebru, T. (2018). Gender shades: Intersectional accuracy disparities in commercial gender classification. In Conference on Fairness, Accountability and Transparency (pp. 77-91).

Burning Glass Technologies. (2016). Beyond point and click: The expanding demand for coding skills. https://www.burning-glass.com/research-project/coding-skills/

Burning Glass Technologies. (2016). The Human Factor: The Hard Time Employers Have Finding Soft Skills. Retrieved from https://www.burning-glass.com/research-project/human-factor/

Businesswire (2020). Tricog Installs AI-Powered ECG Devices in 53 Cities Within 60 Days to Enable Advanced Cardiac Care. https://www.businesswire.com/news/home/20201216005400/en/Tricog-Installs-AI-Powered-ECG-Devices-in-53-Cities-Within-60-Days-to-Enable-Advanced-Cardiac-Care.

Cabrol, N. A. (2018). Alien Mindscapes—A Perspective on the Search for Extraterrestrial Intelligence. Astrobiology, 18(6), 663–676. https://doi.org/10.1089/ast.2017.1664

CAMPBELL, M., HOANE JR, A. J., & HSU, F. H. (2002). DEEP BLUE. ARTIFICIAL
INTELLIGENCE, 134(1-2), 57-83.

CAVOUKIAN, A. (2010). PRIVACY BY DESIGN: THE 7 FOUNDATIONAL PRINCIPLES.
INFORMATION AND PRIVACY COMMISSIONER OF ONTARIO, CANADA.
HTTPS://WWW.IPC.ON.CA/WP-
CONTENT/UPLOADS/RESOURCES/7FOUNDATIONALPRINCIPLES.PDF

CCPA. (2018). CALIFORNIA CONSUMER PRIVACY ACT (CCPA). RETRIEVED FROM
HTTPS://WWW.OAG.CA.GOV/PRIVACY/CCPA

CELLAN-JONES, R. (2014). STEPHEN HAWKING WARNS ARTIFICIAL INTELLIGENCE
COULD END MANKIND. BBC NEWS.
HTTPS://WWW.BBC.COM/NEWS/TECHNOLOGY-30290540

CENTURY TECH (2020). HOW CENTURY WORKS.
HTTPS://WWW.CENTURY.TECH/WHY-CENTURY/.

CHANDLER, A. D. (1977). THE VISIBLE HAND: THE MANAGERIAL REVOLUTION IN
AMERICAN BUSINESS. HARVARD UNIVERSITY PRESS.

CHEN, D., SAIN, S. L., & GUO, K. (2012). DATA MINING FOR THE ONLINE RETAIL
INDUSTRY: A CASE STUDY OF RFM MODEL-BASED CUSTOMER
SEGMENTATION USING DATA MINING. JOURNAL OF DATABASE MARKETING
& CUSTOMER STRATEGY MANAGEMENT, 19(3), 197-208.

CHEN, I. Y., JOHANSSON, F. D., & SONTAG, D. (2018). WHY IS MY CLASSIFIER
DISCRIMINATORY? ADVANCES IN NEURAL INFORMATION PROCESSING
SYSTEMS, 31, 3539–3550.

CHEN, J., LERMAN, K., & FERRARA, E. (2018). TRACKING SOCIAL MEDIA DISCOURSE
ABOUT THE COVID-19 PANDEMIC: DEVELOPMENT OF A PUBLIC
CORONAVIRUS TWITTER DATA SET. JMIR PUBLIC HEALTH AND
SURVEILLANCE, 6(2), E19273.

CHEN, M., MAO, S., & LIU, Y. (2014). BIG DATA: A SURVEY. MOBILE NETWORKS AND
APPLICATIONS, 19(2), 171-209.

CHIEN, S., DOUBLEDAY, J., THOMPSON, D.R. ET AL. (2018). ONBOARD AUTONOMY ON
THE INTELLIGENT PAYLOAD EXPERIMENT CUBESAT MISSION. JOURNAL
OF AEROSPACE INFORMATION SYSTEMS, 15(5), 203-216.
HTTPS://DOI.ORG/10.2514/1.I010515

CHUI, M., MANYIKA, J., MIREMADI, M. (2017). WHAT JOBS WILL BE DONE BY
MACHINES? MCKINSEY GLOBAL INSTITUTE.

CHUI, M., MANYIKA, J., MIREMADI, M., HENKE, N., CHUNG, R., NEL, P., &
MALHOTRA, S. (2018). NOTES FROM THE AI FRONTIER: APPLICATIONS AND

VALUE OF DEEP LEARNING. MCKINSEY GLOBAL INSTITUTE.
HTTPS://WWW.MCKINSEY.COM/~/MEDIA/MCKINSEY/FEATURED%20INSIG
HTS/ARTIFICIAL%20INTELLIGENCE/NOTES%20FROM%20THE%20AI%20FR
ONTIER%20APPLICATIONS%20AND%20VALUE%20OF%20DEEP%20LEARNI
NG/MGI_NOTES-FROM-AI-FRONTIER_DISCUSSION-PAPER.ASHX

COLLINS, F. S., & VARMUS, H. (2015). A NEW INITIATIVE ON PRECISION MEDICINE.
NEW ENGLAND JOURNAL OF MEDICINE, 372(9), 793-795.

COPELAND, B. J. (2006). COLOSSUS: THE SECRETS OF BLETCHLEY PARK'S
CODEBREAKING COMPUTERS. OXFORD UNIVERSITY PRESS.

COSTELLO, J. C., HEISER, L. M., GEORGII, E., GÖNEN, M., MENDEN, M. P., WANG, N.
J., ... & BANSAL, M. (2019). A COMMUNITY EFFORT TO ASSESS AND
IMPROVE DRUG SENSITIVITY PREDICTION ALGORITHMS. NATURE
BIOTECHNOLOGY, 32(12), 1202-1212.

COURSERA. WWW.COURSERA.ORG

COVINGTON, P., ADAMS, J., & SARGIN, E. (2016). DEEP NEURAL NETWORKS FOR
YOUTUBE RECOMMENDATIONS. PROCEEDINGS OF THE 10TH ACM
CONFERENCE ON RECOMMENDER SYSTEMS, 191-198.
HTTPS://DOI.ORG/10.1145/2959100.2959190

CRAWFORD, K. (2021). THE ATLAS OF AI: POWER, POLITICS, AND THE PLANETARY
COSTS OF ARTIFICIAL INTELLIGENCE. YALE UNIVERSITY PRESS.

CUBAN, M. "MARK CUBAN: THE WORLD'S FIRST TRILLIONAIRE WILL BE AN
ARTIFICIAL INTELLIGENCE ENTREPRENEUR." CNBC. (2017).

DARTMOUTH AI CONFERENCE. (1956). RETRIEVED FROM
HTTPS://WWW.AAAI.ORG/ORGANIZATION/DARTMOUTH-
CONFERENCE.PHP

DASTIN, J. (2018, OCTOBER 10). AMAZON SCRAPS SECRET AI RECRUITING TOOL THAT
SHOWED BIAS AGAINST WOMEN. REUTERS. RETRIEVED FROM
HTTPS://WWW.REUTERS.COM/ARTICLE/US-AMAZON-COM-JOBS-
AUTOMATION-INSIGHT/AMAZON-SCRAPS-SECRET-AI-RECRUITING-TOOL-
THAT-SHOWED-BIAS-AGAINST-WOMEN-IDUSKCN1MK08G

DEEPMIND. (2021). ETHICS & SOCIETY. RETRIEVED FROM
HTTPS://DEEPMIND.COM/APPLIED/DEEPMIND-ETHICS-SOCIETY.

DESCARTES, R. (1641). MEDITATIONS ON FIRST PHILOSOPHY.

DEVLIN, J., CHANG, M.-W., LEE, K., & TOUTANOVA, K. (2019). BERT: PRE-TRAINING
OF DEEP BIDIRECTIONAL TRANSFORMERS FOR LANGUAGE
UNDERSTANDING. PROCEEDINGS OF THE 2019 CONFERENCE OF THE

NORTH AMERICAN CHAPTER OF THE ASSOCIATION FOR COMPUTATIONAL
LINGUISTICS: HUMAN LANGUAGE TECHNOLOGIES, VOLUME 1 (LONG AND
SHORT PAPERS), 4171–4186.

DIFTLER, M. A., RADFORD, N. A., MEHLING, J. S., ABDALLAH, M. E., BRIDGWATER, L.
B., SANDERS, A. M., ASKEW, R. S., LINN, D. M., YAMOKOSKI, J. D.,
PERMENTER, F. A., HARGRAVE, B. K., PIATT, B., SAVELY, R. T., &
AMBROSE, R. O. (2011). ROBONAUT 2 - THE FIRST HUMANOID ROBOT IN
SPACE. 2011 IEEE INTERNATIONAL CONFERENCE ON ROBOTICS AND
AUTOMATION, SHANGHAI, CHINA, 2178-2183.
HTTPS://DOI.ORG/10.1109/ICRA.2011.5980382

DONOHO, D. (2017). 50 YEARS OF DATA SCIENCE. JOURNAL OF COMPUTATIONAL AND
GRAPHICAL STATISTICS, 26(4), 745-766.

DWORK, C. (2008). DIFFERENTIAL PRIVACY: A SURVEY OF RESULTS. IN THEORY AND
APPLICATIONS OF MODELS OF COMPUTATION (PP. 1-19). SPRINGER,
BERLIN, HEIDELBERG.

DWORK, C., HARDT, M., PITASSI, T., REINGOLD, O., & ZEMEL, R. (2012). FAIRNESS
THROUGH AWARENESS. IN PROCEEDINGS OF THE 3RD INNOVATIONS IN
THEORETICAL COMPUTER SCIENCE CONFERENCE (PP. 214-226).

EASA. (2015). AUTOMATION IN AVIATION. EUROPEAN UNION AVIATION SAFETY
AGENCY. HTTPS://WWW.EASA.EUROPA.EU/NEWSROOM-AND-
EVENTS/NEWS/AUTOMATION-AVIATION

EDUCATION OUTCOMES FUND (2021). PARTNERING WITH CENTURY TECH TO
IMPROVE LEARNING OUTCOMES THROUGH AI.
HTTPS://EDUCATIONOUTCOMESFUND.ORG/NEWS/PARTNERING-CENTURY-
TECH-IMPROVE-LEARNING-OUTCOMES-THROUGH-AI/.

EDX. WWW.EDX.ORG

ESTEVA, A., KUPREL, B., NOVOA, R. A., KO, J., SWETTER, S. M., BLAU, H. M., &
THRUN, S. (2017). DERMATOLOGIST-LEVEL CLASSIFICATION OF SKIN
CANCER WITH DEEP NEURAL NETWORKS. NATURE, 542(7639), 115-
118.REFERENCES

ETHICALLY ALIGNED DESIGN: A VISION FOR PRIORITIZING HUMAN WELL-BEING
WITH AUTONOMOUS AND INTELLIGENT SYSTEMS. "THE IEEE GLOBAL
INITIATIVE ON ETHICS OF AUTONOMOUS AND INTELLIGENT SYSTEMS.
(2017).

ETHICSNET. (N.D.). RETRIEVED FROM HTTPS://ETHICSNET.ORG/

EU GDPR. (2016). REGULATION (EU) 2016/679 OF THE EUROPEAN PARLIAMENT AND OF THE COUNCIL. RETRIEVED FROM HTTPS://EUR-LEX.EUROPA.EU/ELI/REG/2016/679/OJ

EUCLID. (~300 BC). ELEMENTS.

EUROPEAN COMMISSION (2021). PROPOSAL FOR A REGULATION ON A EUROPEAN APPROACH FOR ARTIFICIAL INTELLIGENCE. EUR-LEX. HTTPS://EUR-LEX.EUROPA.EU/LEGAL-ONTENT/EN/TXT/?URI=CELEX%3A52021PC0206

EUROPEAN UNION. (2018). GENERAL DATA PROTECTION REGULATION (GDPR). RETRIEVED FROM HTTPS://GDPR.EU/

EUROPEAN UNION. (2018). GENERAL DATA PROTECTION REGULATION (GDPR). RETRIEVED FROM HTTPS://EUR-LEX.EUROPA.EU/ELI/REG/2016/679/OJ

FACEBOOK AI (2021). OUR WORK. FACEBOOK AI RESEARCH.

FANNIN, R. (2019). HOW ALIBABA'S AI IS TRANSFORMING THE WAY CHINA SHOPS ONLINE. FORBES. HTTPS://WWW.FORBES.COM/SITES/REBECCAFANNIN/2019/07/25/HOW-ALIBABAS-AI-IS-TRANSFORMING-THE-WAY-CHINA-SHOPS-ONLINE/?SH=59C2203424FB

FARMDRIVE (2021). FARMDRIVE: ENABLING FINANCIAL ACCESS FOR SMALLHOLDER FARMERS. HTTPS://WWW.FARMDRIVE.CO.KE/

FERRUCCI, D. (2012). INTRODUCTION TO "THIS IS WATSON". IBM JOURNAL OF RESEARCH AND DEVELOPMENT, 56(3.4), 1:1–1:15.

FERRUCCI, D., LEVAS, A., BAGCHI, S., GONDEK, D., & MUELLER, E. T. (2013). WATSON: BEYOND JEOPARDY!. ARTIFICIAL INTELLIGENCE, 199-200, 93-105.

FERRY, Q., STEINBERG, J., WEBBER, C., FITZPATRICK, D. R., PONTING, C. P., ZISSERMAN, A., & NELLÅKER, C. (2020). DIAGNOSTICALLY RELEVANT FACIAL GESTALT INFORMATION FROM ORDINARY PHOTOS. ELIFE, 9, E54075

FLORIDI, L., & COWLS, J. (2019). A UNIFIED FRAMEWORK OF FIVE PRINCIPLES FOR AI IN SOCIETY. HARVARD DATA SCIENCE REVIEW, 1(1).

FONG, T., BUALAT, M., DEANS, M., & HEGGY, E. (2019). AI IN SPACE EXPLORATION. AI MAGAZINE, 40(2), 14-24. HTTPS://DOI.ORG/10.1609/AIMAG.V40I2.2850

FORSYTH, D. A., & PONCE, J. (2011). COMPUTER VISION: A MODERN APPROACH. PEARSON.

FRANKLIN, B. "AN INVESTMENT IN KNOWLEDGE ALWAYS PAYS THE BEST INTEREST." GOODREADS. (N.D.).

FREY, C. B. (2019). THE TECHNOLOGY TRAP: CAPITAL, LABOR, AND POWER IN THE
AGE OF AUTOMATION. PRINCETON UNIVERSITY PRESS.

FREY, C. B., & OSBORNE, M. A. (2017). THE FUTURE OF EMPLOYMENT: HOW
SUSCEPTIBLE ARE JOBS TO COMPUTERIZATION? TECHNOLOGICAL
FORECASTING AND SOCIAL CHANGE, 114, 254-280.

FUJIMOTO, K., HANADA, T., & HIROSE, T. (2020). ARTIFICIAL INTELLIGENCE FOR
SPACE TRAFFIC MANAGEMENT. 2020 59TH ANNUAL CONFERENCE OF THE
SOCIETY OF INSTRUMENT AND CONTROL ENGINEERS OF JAPAN (SICE),
CHIBA, JAPAN, 1336-1340.
HTTPS://DOI.ORG/10.23919/SICE48014.2020.9242362

FURMAN, J., SEAMANS, R., & KOMINERS, S. D. (2019). A PROPOSAL FOR
MODERNIZING LABOR LAWS FOR TWENTY-FIRST-CENTURY WORK: THE
INDEPENDENT WORKER. THE HAMILTON PROJECT, POLICY PROPOSAL
2019-08. RETRIEVED FROM
HTTPS://WWW.HAMILTONPROJECT.ORG/ASSETS/FILES/MODERNIZING_LA
BOR_LAWS_FOR_TWENTY_FIRST_CENTURY_WORK.PDF.

FUTOMA, J., SIMONS, M., PANCH, T., DOSHI-VELEZ, F., & CELI, L. A. (2020). THE
MYTH OF GENERALIZABILITY IN CLINICAL RESEARCH AND MACHINE
LEARNING IN HEALTH CARE. THE LANCET DIGITAL HEALTH, 2(9), E489-
E492.

FUTURE OF LIFE INSTITUTE. (2015). AUTONOMOUS WEAPONS: AN OPEN LETTER
FROM AI & ROBOTICS RESEARCHERS. RETRIEVED FROM
HTTPS://FUTUREOFLIFE.ORG/AUTONOMOUS-WEAPONS-OPEN-LETTER-2015

FUTURE OF LIFE INSTITUTE. (N.D.). AI ETHICS READING LIST. RETRIEVED FROM
HTTPS://FUTUREOFLIFE.ORG/AI-ETHICS-RESOURCES/

GEBRU, T. (2021). DOCUMENTING, CRITIQUING, AND SHAPING AI. PROCEEDINGS OF
THE AAAI CONFERENCE ON ARTIFICIAL INTELLIGENCE, 35(18), 15806-
15810.

GOERTZEL, B., & PENNACHIN, C. (EDS.). (2007). ARTIFICIAL GENERAL
INTELLIGENCE (VOL. 2). SPRINGER.

GOLDBERG, Y. (2017). NEURAL NETWORK METHODS FOR NATURAL LANGUAGE
PROCESSING. MORGAN & CLAYPOOL PUBLISHERS.

GOLDSTINE, H. H. (1972). THE COMPUTER FROM PASCAL TO VON NEUMANN.
PRINCETON UNIVERSITY PRESS.

GOMEZ-URIBE, C. A., & HUNT, N. (2015). THE NETFLIX RECOMMENDER SYSTEM:
ALGORITHMS, BUSINESS VALUE, AND INNOVATION. ACM TRANSACTIONS

ON MANAGEMENT INFORMATION SYSTEMS, 6(4), 1–19. DOI: 10.1145/2843948

GOMEZ-URIBE, C. A., & HUNT, N. (2015). THE NETFLIX RECOMMENDER SYSTEM: ALGORITHMS, BUSINESS VALUE, AND INNOVATION. ACM TRANSACTIONS ON MANAGEMENT INFORMATION SYSTEMS (TMIS), 6(4), 1-19.

GOMEZ-URIBE, C. A., & HUNT, N. (2016). THE NETFLIX RECOMMENDER SYSTEM: ALGORITHMS, BUSINESS VALUE, AND INNOVATION. ACM TRANSACTIONS ON MANAGEMENT INFORMATION SYSTEMS (TMIS), 6(4), 1-19.

GONZALEZ, R. C., & WOODS, R. E. (2017). DIGITAL IMAGE PROCESSING. PEARSON.

GOODFELLOW, I. J., SHLENS, J., & SZEGEDY, C. (2015). EXPLAINING AND HARNESSING ADVERSARIAL EXAMPLES. IN PROCEEDINGS OF THE INTERNATIONAL CONFERENCE ON LEARNING REPRESENTATIONS (ICLR).

GOODFELLOW, I., BENGIO, Y., & COURVILLE, A. (2016). DEEP LEARNING. MIT PRESS.

GOOGLE AI. (2019). USING AI TO PREDICT FLOOD RISK. HTTPS://AI.GOOGLEBLOG.COM/2019/06/USING-AI-TO-PREDICT-FLOODS-AND-HELP.HTML.

GOOGLE AI. (2021). MAKING GOOGLE PHOTOS BETTER WITH MACHINE LEARNING. RETRIEVED FROM HTTPS://AI.GOOGLEBLOG.COM/2021/06/MAKING-GOOGLE-PHOTOS-BETTER-WITH.HTML

GOOGLE. (2018). GOOGLE AI PRINCIPLES. HTTPS://AI.GOOGLE/PRINCIPLES/

GROTHER, P., NGAN, M., & HANAOKA, K. (2019). FACE RECOGNITION VENDOR TEST (FRVT) PART 3: DEMOGRAPHIC EFFECTS. NATIONAL INSTITUTE OF STANDARDS AND TECHNOLOGY. HTTPS://NVLPUBS.NIST.GOV/NISTPUBS/IR/2019/NIST.IR.8280.PDF

GRUBER, T. R. (1993). A TRANSLATION APPROACH TO PORTABLE ONTOLOGY SPECIFICATIONS. KNOWLEDGE ACQUISITION, 5(2), 199-220.

GUALTIERI, M. (2017). "THE FORRESTER WAVE: COGNITIVE SEARCH AND KNOWLEDGE DISCOVERY SOLUTIONS." FORRESTER RESEARCH.

GULSHAN, V., PENG, L., CORAM, M., STUMPE, M. C., WU, D., NARAYANASWAMY, A., … & WEBSTER, D. R. (2016). DEVELOPMENT AND VALIDATION OF A DEEP LEARNING ALGORITHM FOR DETECTION OF DIABETIC RETINOPATHY IN RETINAL FUNDUS PHOTOGRAPHS. JAMA, 316(22), 2402–2410. RETAIL – AMAZON

GUZELLA, T. S., & CAMINHAS, W. M. (2009). A REVIEW OF MACHINE LEARNING APPROACHES TO SPAM FILTERING. EXPERT SYSTEMS WITH APPLICATIONS, 36(7), 10206-10222.

HAJIAN, S., BONCHI, F., & CASTILLO, C. (2016). *ALGORITHMIC BIAS: FROM DISCRIMINATION DISCOVERY TO FAIRNESS-AWARE DATA MINING. PROCEEDINGS OF THE 22ND ACM SIGKDD INTERNATIONAL CONFERENCE ON KNOWLEDGE DISCOVERY AND DATA MINING*, 2125-2126.

HAN, J., PEI, J., & KAMBER, M. (2011). *DATA MINING: CONCEPTS AND TECHNIQUES* (3RD ED.). MORGAN KAUFMANN.

HARNAD, S. (2000). *TURING INDISTINGUISHABILITY AND THE BLIND WATCHMAKER. IN MACHINES AND THOUGHT: THE LEGACY OF ALAN TURING* (VOL. 1, PP. 3-12). OXFORD UNIVERSITY PRESS.

HASTIE, T., TIBSHIRANI, R., & FRIEDMAN, J. (2009). *THE ELEMENTS OF STATISTICAL LEARNING: DATA MINING, INFERENCE, AND PREDICTION.* SPRINGER.

HAWKINS, A. (2021). *NTSB: TESLA'S AUTOPILOT SYSTEM AND DISTRACTED DRIVER TO BLAME FOR FATAL CRASH. THE VERGE.*
HTTPS://WWW.THEVERGE.COM/2021/5/29/22458509/NTSB-TESLA-AUTOPILOT-FATAL-CRASH-CAUSE

HAYES, P. J., & MICHIE, D. (1985). "THE SECOND AI WINTER." *AI MAGAZINE*, 6(2), 59-64.

HAYES-ROTH, B., & WATERMAN, D. A. (1990). *PATTERN-DIRECTED INFERENCE SYSTEMS. IN READINGS IN ARTIFICIAL INTELLIGENCE* (PP. 97-105). MORGAN KAUFMANN.

HEATH, T. L. (1921). *A HISTORY OF GREEK MATHEMATICS.* CLARENDON PRESS.

HILL, K. (2020). *WRONGFULLY ACCUSED BY AN ALGORITHM. THE NEW YORK TIMES.*
HTTPS://WWW.NYTIMES.COM/2020/06/24/TECHNOLOGY/FACIAL-RECOGNITION-ARREST.HTML

HIRSCHBERG, J., & MANNING, C. D. (2015). *ADVANCES IN NATURAL LANGUAGE PROCESSING. SCIENCE*, 349(6245), 261–266. DOI: 10.1126/SCIENCE.AAA8685

HOLLISTER, S. (2019). *ELON MUSK: TESLA'S NEW SELF-DRIVING CHIP IS FINALLY BFFS WITH THE AUTOPILOT AI. THE VERGE.*
HTTPS://WWW.THEVERGE.COM/2019/4/22/18510897/ELON-MUSK-TESLA-AUTOPILOT-FULL-SELF-DRIVING-AI-DAY

HOROWITZ, M. C., & SCHARRE, P. (2015). *AN INTRODUCTION TO AUTONOMY IN WEAPON SYSTEMS.* CENTER FOR A NEW AMERICAN SECURITY.

HOY, M. B. (2018). ALEXA, SIRI, CORTANA, AND MORE: AN INTRODUCTION TO VOICE ASSISTANTS. MEDICAL REFERENCE SERVICES QUARTERLY, 37(1), 81–88. DOI: 10.1080/02763869.2018.1404391

HUANG, M. H., & RUST, R. T. (2018). ARTIFICIAL INTELLIGENCE IN SERVICE. JOURNAL OF SERVICE RESEARCH, 21(2), 155-172.

HUTCHINS, W. J. (2018). THE ORIGINS OF MACHINE TRANSLATION. IN THE ROUTLEDGE HANDBOOK OF TRANSLATION STUDIES AND LINGUISTICS.

IBM RESEARCH. (1997). DEEP BLUE. RETRIEVED FROM HTTPS://WWW.RESEARCH.IBM.COM/DEEPBLUE/MEET/HTML/D.3.SHTML

IBM WATSON. (N.D.). "WHAT IS COGNITIVE COMPUTING?" RETRIEVED FROM HTTPS://WWW.IBM.COM/WATSON/COGNITIVE-COMPUTING

INDEX OF DIGITAL INCLUSIVENESS." WORLD ECONOMIC FORUM. (2019).

INTERNATIONAL TELECOMMUNICATION UNION (ITU). (2020). MEASURING DIGITAL DEVELOPMENT: FACTS AND FIGURES 2020. ITU. HTTPS://WWW.ITU.INT/EN/ITU-D/STATISTICS/PAGES/FACTS/DEFAULT.ASPX

JACKSON, P. (1999). INTRODUCTION TO EXPERT SYSTEMS. ADDISON-WESLEY.

JAGTIANI, J., & LEMIEUX, C. (2017). THE ROLES OF ALTERNATIVE DATA AND MACHINE LEARNING IN FINTECH LENDING: EVIDENCE FROM THE LENDINGCLUB CONSUMER PLATFORM. FEDERAL RESERVE BANK OF PHILADELPHIA, RESEARCH DEPARTMENT, WORKING PAPERS, 17-17.

JAQUET-DROZ, P., JAQUET-DROZ, J.-F., & CHAPUIS, A. (2018). JAQUET-DROZ AND HIS MECHANICAL ANDROIDS: THE WRITER, THE DRAFTSMAN, AND THE MUSICIAN. SPRINGER.

JEAN, N., BURKE, M., XIE, M., DAVIS, W. M., LOBELL, D. B., & ERMON, S. (2016). COMBINING SATELLITE IMAGERY AND MACHINE LEARNING TO PREDICT POVERTY. SCIENCE, 353(6301), 790-794. HTTPS://DOI.ORG/10.1126/SCIENCE.AAF7894

JOBIN, A., IENCA, M., & VAYENA, E. (2019). THE GLOBAL LANDSCAPE OF AI ETHICS GUIDELINES. NATURE MACHINE INTELLIGENCE, 1(9), 389-399. HTTPS://DOI.ORG/10.1038/S42256-019-0088-2

JURAFSKY, D., & MARTIN, J. H. (2019). SPEECH AND LANGUAGE PROCESSING (3RD ED.). PEARSON.

JURAFSKY, D., & MARTIN, J. H. (2020). SPEECH AND LANGUAGE PROCESSING. RETRIEVED FROM HTTPS://WEB.STANFORD.EDU/~JURAFSKY/SLP3/

Kaelbling, L. P., Littman, M. L., & Moore, A. W. (1996). Reinforcement learning: A survey. Journal of Artificial Intelligence Research, 4, 237-285.

Kania, E. (2021). Military applications of artificial intelligence: Ethical concerns in an uncertain world. Ethics & International Affairs, 35(1), 67-87.

Kania, E. (2021). South Korea's AI Capabilities and Diplomacy. Center for a New American Security. Retrieved from https://www.cnas.org/publications/reports/south-koreas-ai-capabilities-and-diplomacy

Kannan, A., Kurach, K., Ravi, S., Kaufmann, T., Tomkins, A., Miklos, B., ... & Ganea, O. (2016). Smart Reply: Automated Response Suggestion for Email. In Proceedings of the 22nd ACM SIGKDD International Conference on Knowledge Discovery and Data Mining (pp. 955-964).

Karanja, S. (2018). Kenya's Ministry of ICT Sets up Blockchain and Artificial Intelligence Taskforce. BitcoinAfrica.io. https://bitcoinafrica.io/2018/02/28/kenya-blockchain-taskforce/

Kelly, K. (2016). The Inevitable: Understanding the 12 Technological Forces That Will Shape Our Future. Penguin Books.

Kootstra, G. (2020). The Robots Are Coming: A Guide to the Future of Artificial Intelligence. Bloomsbury Publishing.

Koren, Y., Bell, R., & Volinsky, C. (2009). Matrix factorization techniques for recommender systems. Computer, (8), 30-37.

Krizhevsky, A., Sutskever, I., & Hinton, G. E. (2012). Imagenet classification with deep convolutional neural networks. In Advances in neural information processing systems (pp. 1097-1105).

Kurzweil, R. (2005). The Singularity Is Near: When Humans Transcend Biology. Viking.

Kusner, M. J., Loftus, J., Russell, C., & Silva, R. (2017). Counterfactual Fairness. In Proceedings of the 34th International Conference on Machine Learning (Vol. 70, pp. 2137–2145).

Landes, D. S. (1983). Revolution in Time: Clocks and the Making of the Modern World. Harvard University Press.

LaValle, S. M. (2006). Planning Algorithms. Cambridge University Press.

LAVINGTON, S. (2014). "ALAN TURING AND HIS CONTEMPORARIES: BUILDING THE
WORLD'S FIRST COMPUTERS." THE MIT PRESS.

LECUN, Y., BENGIO, Y., & HINTON, G. (2015). DEEP LEARNING. NATURE, 521(7553),
436-444.

LEE, K. F. (2018). AI SUPERPOWERS: CHINA, SILICON VALLEY, AND THE NEW WORLD
ORDER. HOUGHTON MIFFLIN HARCOURT.

LEE, K.-F. (2018). AI SUPERPOWERS: CHINA, SILICON VALLEY, AND THE NEW WORLD
ORDER. HOUGHTON MIFFLIN HARCOURT.

LEIBNIZ, G. W. (1680). DISSERTATIO DE ARTE COMBINATORIA (DISSERTATION ON
THE ART OF COMBINATIONS).

LI, F. (2018). HOW WE'RE TEACHING COMPUTERS TO UNDERSTAND PICTURES. TED
TALK.

LI, X., & KARAHANNA, E. (2015). ONLINE RECOMMENDATION SYSTEMS IN A B2C E-
COMMERCE CONTEXT: A REVIEW AND FUTURE DIRECTIONS. JOURNAL OF
THE ASSOCIATION FOR INFORMATION SYSTEMS, 16(2), 72-107.

LI, X., DUNN, J., SALINS, D., ET AL. (2020). DIGITAL HEALTH: TRACKING PHYSIOMES
AND ACTIVITY USING WEARABLE BIOSENSORS REVEALS USEFUL HEALTH-
RELATED INFORMATION. PLOS BIOLOGY.

LINKEDIN (2020). 2020 EMERGING JOBS REPORT.
HTTPS://BUSINESS.LINKEDIN.COM/TALENT-SOLUTIONS/BLOG/TRENDS-
AND-RESEARCH/2020/EMERGING-JOBS-REPORT

LINO, C., CHRISTIE, M., & THALMANN, D. (2014). THE DIRECTOR'S LENS: AN
INTELLIGENT ASSISTANT FOR VIRTUAL CINEMATOGRAPHY. IN
PROCEEDINGS OF THE 19TH ACM SYMPOSIUM ON VIRTUAL REALITY
SOFTWARE AND TECHNOLOGY - VRST '13 (P. 167). ACM PRESS. DOI:
10.1145/2503713.2503715

LUCKIN, R. (2018). MACHINE LEARNING AND HUMAN INTELLIGENCE: THE FUTURE
OF EDUCATION FOR THE 21ST CENTURY. BRITISH JOURNAL OF
EDUCATIONAL STUDIES, 66(3), 302-319.

MAEDCHE, A., & STAAB, S. (2001). ONTOLOGY LEARNING FOR THE SEMANTIC WEB.
IEEE INTELLIGENT SYSTEMS, 16(2), 72-79.

MANNING, C. D., & SCHÜTZE, H. (1999). FOUNDATIONS OF STATISTICAL NATURAL
LANGUAGE PROCESSING. MIT PRESS.

MARCUS, G., & DAVIS, E. (2019). REBOOTING AI: BUILDING ARTIFICIAL
INTELLIGENCE WE CAN TRUST. PANTHEON BOOKS.

MARR, B. (2018). HOW MUCH DATA DO WE CREATE EVERY DAY? THE MIND-
BLOWING STATS EVERYONE SHOULD READ. RETRIEVED FROM:
HTTPS://WWW.FORBES.COM/SITES/BERNARDMARR/2018/05/21/HOW-
MUCH-DATA-DO-WE-CREATE-EVERY-DAY-THE-MIND-BLOWING-STATS-
EVERYONE-SHOULD-READ/

MASSACHUSETTS INSTITUTE OF TECHNOLOGY. (N.D.). ARTIFICIAL INTELLIGENCE
ETHICS AND PRACTICE. RETRIEVED FROM
HTTPS://WWW.PROFESSIONALEDUCATIONPE.MIT.EDU/COURSES/ETHICS-
OF-ARTIFICIAL-INTELLIGENCE

MCCARTHY, J. (1956). PROPOSAL FOR THE DARTMOUTH SUMMER RESEARCH
PROJECT ON ARTIFICIAL INTELLIGENCE. RETRIEVED FROM
HTTPS://WWW-
FORMAL.STANFORD.EDU/JMC/HISTORY/DARTMOUTH/DARTMOUTH.HTML

MCCARTHY, J. (2007). WHAT IS ARTIFICIAL INTELLIGENCE? STANFORD UNIVERSITY.
RETRIEVED FROM: HTTP://JMC.STANFORD.EDU/ARTIFICIAL-
INTELLIGENCE/WHAT-IS-AI/INDEX.HTML

MCCARTHY, J., MINSKY, M. L., ROCHESTER, N., & SHANNON, C. E. (1955). A
PROPOSAL FOR THE DARTMOUTH SUMMER RESEARCH PROJECT ON
ARTIFICIAL INTELLIGENCE. RETRIEVED FROM
HTTPS://WWW.DARTMOUTH.EDU/~AI50/DOCUMENTS/SHORT-MCCARTHY-
DARTMOUTH-AI.PDF

MCCARTHY, J., MINSKY, M. L., ROCHESTER, N., & SHANNON, C. E. (1955). "A
PROPOSAL FOR THE DARTMOUTH SUMMER RESEARCH PROJECT ON
ARTIFICIAL INTELLIGENCE." AI MAGAZINE, 27(4), 12-14.

MCKENZIE, J. (2013). HOW RETAILERS CAN KEEP UP WITH CONSUMERS. MCKINSEY
& COMPANY.

MCKINNEY, W. (2011). PANDAS: A FOUNDATIONAL PYTHON LIBRARY FOR DATA
ANALYSIS AND STATISTICS. PYTHON FOR HIGH PERFORMANCE AND
SCIENTIFIC COMPUTING, 14.

MICROSOFT (2020). WRITE WITH CONFIDENCE ACROSS DOCUMENTS, EMAIL, AND
THE WEB WITH FEATURES THAT HELP STRENGTHEN YOUR SPELLING,
GRAMMAR, AND STYLE. MICROSOFT EDITOR.

MICROSOFT AI (2021). EMPOWERMENT BEGINS WITH INCLUSION. AI FOR HEALTH.

MICROSOFT AZURE (2021). AZURE AI—ARTIFICIAL INTELLIGENCE APPS & AGENTS.
MICROSOFT AZURE.

MIKOLOV, T., SUTSKEVER, I., CHEN, K., CORRADO, G. S., & DEAN, J. (2013). DISTRIBUTED REPRESENTATIONS OF WORDS AND PHRASES AND THEIR COMPOSITIONALITY. IN ADVANCES IN NEURAL INFORMATION PROCESSING SYSTEMS (PP. 3111-3119).

MINSKY, M. (1968). SEMANTIC INFORMATION PROCESSING. MIT PRESS.

MISHRA, D. (2018). HOW AI IS DRIVING A FAST AND FURIOUS REVOLUTION. ECONOMIC TIMES. HTTPS://ECONOMICTIMES.INDIATIMES.COM/TECH/INTERNET/HOW-AI-IS-DRIVING-A-FAST-AND-FURIOUS-REVOLUTION/ARTICLESHOW/64134888.CMS

MITCHELL, T. M. (1997). MACHINE LEARNING. MCGRAW-HILL.

MITTELSTADT, B. (2019). PRINCIPLES ALONE CANNOT GUARANTEE ETHICAL AI. NATURE MACHINE INTELLIGENCE, 1(11), 501-507.

MNIH, V., KAVUKCUOGLU, K., SILVER, D., ET AL. (2015). HUMAN-LEVEL CONTROL THROUGH DEEP REINFORCEMENT LEARNING. NATURE, 518(7540), 529-533.

MORGAN, S. (2017). CYBERSECURITY LABOR CRUNCH TO HIT 3.5 MILLION UNFILLED JOBS BY 2021. CYBERCRIME MAGAZINE.

MULUPI, D. (2016). KENYAN AGRITECH STARTUP UJUZIKILIMO USES SENSORS TO HELP FARMERS. HOW WE MADE IT IN AFRICA. HTTPS://WWW.HOWWEMADEITINAFRICA.COM/KENYAN-AGRI-TECH-STARTUP-UJUZIKILIMO-USES-SENSORS-HELP-FARMERS/53456/

MURRAY, C. (1986). HERO OF ALEXANDRIA. JOHNS HOPKINS UNIVERSITY PRESS.

MUSK, E. (2014). ELON MUSK: ARTIFICIAL INTELLIGENCE IS OUR BIGGEST EXISTENTIAL THREAT. THE GUARDIAN. HTTPS://WWW.THEGUARDIAN.COM/TECHNOLOGY/2014/OCT/27/ELON-MUSK-ARTIFICIAL-INTELLIGENCE-AI-BIGGEST-EXISTENTIAL-THREAT

MUSK, E. (2014, OCTOBER 24). ELON MUSK: 'WITH ARTIFICIAL INTELLIGENCE WE ARE SUMMONING THE DEMON.' THE GUARDIAN. RETRIEVED FROM: HTTPS://WWW.THEGUARDIAN.COM/TECHNOLOGY/2014/OCT/27/ELON-MUSK-ARTIFICIAL-INTELLIGENCE-AI-BIGGEST-EXISTENTIAL-THREAT

NAMBIAR, S. (2018). HOW THIS AGRITECH FIRM IS USING AI AND SATELLITE IMAGERY TO MONITOR CROPS. ANALYTICS INDIA MAGAZINE. HTTPS://ANALYTICSINDIAMAG.COM/HOW-THIS-AGRITECH-FIRM-IS-USING-AI-AND-SATELLITE-IMAGERY-TO-MONITOR-CROPS/

NARAYANAN, A., & SHMATIKOV, V. (2010). DE-ANONYMIZING SOCIAL NETWORKS. 2009 30TH IEEE SYMPOSIUM ON SECURITY AND PRIVACY. HTTPS://DOI.ORG/10.1109/SP.2009.22

NASA. (2021). MARS 2020 PERSEVERANCE ROVER. HTTPS://WWW.NASA.GOV/PERSEVERANCE

NATIONAL TRANSPORTATION SAFETY BOARD (2019). COLLISION BETWEEN VEHICLE CONTROLLED BY DEVELOPMENTAL AUTOMATED DRIVING SYSTEM AND PEDESTRIAN. HTTPS://WWW.NTSB.GOV/NEWS/PRESS-RELEASES/PAGES/NR20191119C.ASPX

NAUGHTON, K. (2021). TESLA'S AUTOPILOT HAS COMPETITION IN PUSH TO BRING SELF-DRIVING CARS TO MARKET. BLOOMBERG. HTTPS://WWW.BLOOMBERG.COM/NEWS/ARTICLES/2021-08-19/TESLA-S-TSLA-AUTOPILOT-FACES-COMPETITION-AS-SELF-DRIVING-CARS-NEAR

NEWELL, A., & SIMON, H. A. (1956). "THE LOGIC THEORIST—AN APPROACH TO MACHINE LEARNING." JOURNAL OF COUNSELING PSYCHOLOGY, 2(1), 1-30.

NEWELL, A., SHAW, J. C., & SIMON, H. A. (1959). "REPORT ON A GENERAL PROBLEM-SOLVING PROGRAM." PROCEEDINGS OF THE INTERNATIONAL CONFERENCE ON INFORMATION PROCESSING, UNESCO, PARIS, 256-264.

NG, A. "AI TRANSFORMATION PLAYBOOK." LANDING AI. (2018).

NG, A. (2016). WHAT ARTIFICIAL INTELLIGENCE CAN AND CAN'T DO RIGHT NOW. HARVARD BUSINESS REVIEW. RETRIEVED FROM: HTTPS://HBR.ORG/2016/11/WHAT-ARTIFICIAL-INTELLIGENCE-CAN-AND-CANT-DO-RIGHT-NOW

NG, A. (2017). AI IS THE NEW ELECTRICITY. STANFORD BUSINESS.

NG, A. (2019). AI FOR EVERYONE. COURSERA. HTTPS://WWW.COURSERA.ORG/LEARN/AI-FOR-EVERYONE

NIELSEN, M. (2015). NEURAL NETWORKS AND DEEP LEARNING. DETERMINATION PRESS.

NILSSON, N. J. (1998). "ARTIFICIAL INTELLIGENCE: A PERSONAL VIEW." AI MAGAZINE, 19(2), 13-14.

NILSSON, N. J. (1998). ARTIFICIAL INTELLIGENCE: A NEW SYNTHESIS. MORGAN KAUFMANN.

NILSSON, N. J. (2010). THE QUEST FOR ARTIFICIAL INTELLIGENCE: A HISTORY OF IDEAS AND ACHIEVEMENTS. CAMBRIDGE: CAMBRIDGE UNIVERSITY PRESS.

NOFFZ, K. H., SRINIVASAN, K., & DELAUNE, R. A. (2019). USING MACHINE LEARNING TO ANALYZE AND OPTIMIZE DEEP SPACE COMMUNICATIONS NETWORKS. 2019 IEEE Aerospace Conference, BIG SKY, MT, USA, 1-13. HTTPS://DOI.ORG/10.1109/AERO.2019.8741744

NORVIG, P., & RUSSELL, S. (2012). ARTIFICIAL INTELLIGENCE: A MODERN APPROACH. PEARSON.

NOY, N. F., & MCGUINNESS, D. L. (2001). ONTOLOGY DEVELOPMENT 101: A GUIDE TO CREATING YOUR FIRST ONTOLOGY. STANFORD KNOWLEDGE SYSTEMS LABORATORY TECHNICAL REPORT KSL-01-05 AND STANFORD MEDICAL INFORMATICS TECHNICAL REPORT SMI-2001-0880.

OBERMEYER, Z., POWERS, B., VOGELI, C., & MULLAINATHAN, S. (2019). DISSECTING RACIAL BIAS IN AN ALGORITHM USED TO MANAGE THE HEALTH OF POPULATIONS. SCIENCE, 366(6464), 447-453.

O'DELL, J. (2011). THE INFINITE ART OF MINECRAFT: WHY THE GAME IS A BIG DEAL. MASHABLE. RETRIEVED FROM: HTTPS://MASHABLE.COM/2011/11/07/MINECRAFT/

OECD (2019). RECOMMENDATION OF THE COUNCIL ON ARTIFICIAL INTELLIGENCE. OECD LEGAL INSTRUMENTS. HTTPS://LEGALINSTRUMENTS.OECD.ORG/EN/INSTRUMENTS/OECD-LEGAL-0449

OECD (2020). BRIDGING THE DIGITAL DIVIDE IN DEVELOPING COUNTRIES. OECD. HTTP://WWW.OECD.ORG/STI/BRIDGING-THE-DIGITAL-DIVIDE-IN-DEVELOPING-COUNTRIES-248AFBA0-EN.HTML

OECD. (2019). AI PRINCIPLES. RETRIEVED FROM HTTP://WWW.OECD.ORG/GOING-DIGITAL/AI/PRINCIPLES/

O'NEIL, C. (2016). WEAPONS OF MATH DESTRUCTION: HOW BIG DATA INCREASES INEQUALITY AND THREATENS DEMOCRACY. CROWN.

OPENAI. (2018). OPENAI FIVE. RETRIEVED FROM: HTTPS://OPENAI.COM/RESEARCH/OPENAI-FIVE/

OPENAI. (2021). OPENAI CHARTER. RETRIEVED FROM HTTPS://OPENAI.COM/CHARTER/

OPENAI. (N.D.). ETHICS RESOURCES. RETRIEVED FROM HTTPS://WWW.OPENAI.COM/RESOURCES/ETHICS-RESOURCES/

OVID. (8 AD). METAMORPHOSES. PENGUIN CLASSICS.

PARISER, E. (2011). THE FILTER BUBBLE: WHAT THE INTERNET IS HIDING FROM YOU. PENGUIN PRESS.

PARTNERSHIP ON AI. (2021). PARTNERSHIP ON AI.

HTTPS://WWW.PARTNERSHIPONAI.ORG/

PARTNERSHIP ON AI. (2021). PARTNERSHIP ON AI'S TENETS. RETRIEVED FROM

HTTPS://WWW.PARTNERSHIPONAI.ORG/ABOUT/#TENETS

PARTNERSHIP ON AI. (2021). RETRIEVED FROM

HTTPS://WWW.PARTNERSHIPONAI.ORG/

PARTNERSHIP ON AI. (N.D.). ABOUT US. RETRIEVED FROM

HTTPS://WWW.PARTNERSHIPONAI.ORG/ABOUT/

PASQUALE, F. (2015). THE BLACK BOX SOCIETY: THE SECRET ALGORITHMS THAT
CONTROL MONEY AND INFORMATION. HARVARD UNIVERSITY PRESS.

PATEL, J., SHAH, S., THAKKAR, P., & KOTECHA, K. (2015). PREDICTING STOCK AND
STOCK PRICE INDEX MOVEMENT USING TREND DETERMINISTIC DATA
PREPARATION AND MACHINE LEARNING TECHNIQUES. EXPERT SYSTEMS
WITH APPLICATIONS, 42(1), 259-268.

PAYNE, K. (2019). THE MUSE IS IN THE MACHINE: CREATIVITY, NOVELTY, AND AI
MUSIC COMPOSERS. PERSPECTIVES ON SCIENCE, 27(4), 569–591. DOI:
10.1162/POSC_A_00312

PEARSON, K. A., FEINSTEIN, A. D., & DE LEE, N. (2019). SEARCHING FOR
EXOPLANETS USING ARTIFICIAL INTELLIGENCE. MONTHLY NOTICES OF
THE ROYAL ASTRONOMICAL SOCIETY, 485(3), 3499-3507.

HTTPS://DOI.ORG/10.1093/MNRAS/STZ509

PEDREGOSA, F., VAROQUAUX, G., GRAMFORT, A., MICHEL, V., THIRION, B., GRISEL,
O., BLONDEL, M., PRETTENHOFER, P., WEISS, R., DUBOURG, V.,
VANDERPLAS, J., PASSOS, A., COURNAPEAU, D., BRUCHER, M., PERROT,
M., & DUCHESNAY, E. (2011). SCIKIT-LEARN: MACHINE LEARNING IN
PYTHON. JOURNAL OF MACHINE LEARNING RESEARCH, 12, 2825-2830.

PROFESSIONAL CERTIFICATE IN AI AND MACHINE LEARNING. EDX. RETRIEVED
FROM HTTPS://WWW.EDX.ORG/PROFESSIONAL-CERTIFICATE/AI-AND-
MACHINE-LEARNING

PROVOST, F., & FAWCETT, T. (2013). DATA SCIENCE FOR BUSINESS: WHAT YOU NEED
TO KNOW ABOUT DATA MINING AND DATA-ANALYTIC THINKING. O'REILLY
MEDIA, INC.

PYTHON FOR DATA SCIENCE AND MACHINE LEARNING BOOTCAMP. UDEMY.
RETRIEVED FROM HTTPS://WWW.UDEMY.COM/COURSE/PYTHON-FOR-
DATA-SCIENCE-AND-MACHINE-LEARNING-BOOTCAMP/

Raghupathi, W., & Raghupathi, V. (2014). Big data analytics in healthcare: promise and potential. Health Information Science and Systems, 2(1), 3.

Rahwan, I., Cebrian, M., Obradovich, N., Bongard, J., Bonnefon, J. F., Breazeal, C., … & Jennings, N. R. (2019). Machine behavior. Nature, 568(7753), 477-486.

Rajkomar, A., Dean, J., & Kohane, I. (2019). Machine Learning in Medicine. New England Journal of Medicine, 380, 1347-1358.

Rajkomar, A., Oren, E., Chen, K., Dai, A. M., Hajaj, N., Hardt, M., … & Sun, M. (2018). Scalable and accurate deep learning with electronic health records. NPJ Digital Medicine, 1(1), 1-10.

Rauschnabel, P. A., Rossmann, A., & tom Dieck, M. C. (2018). An adoption framework for mobile augmented reality games: The case of Pokémon Go. Computers in Human Behavior, 76, 276-286.

Ricci, F., Rokach, L., & Shapira, B. (2011). Introduction to Recommender Systems Handbook. Springer.

Ricci, F., Rokach, L., & Shapira, B. (2015). Recommender Systems: Introduction and Challenges. In Recommender Systems Handbook (pp. 1–34). Springer US. doi: 10.1007/978-1-4899-7637-6_1

Ricci, F., Rokach, L., & Shapira, B. (2015). Recommender Systems: Introduction and Challenges. In Recommender Systems Handbook (pp. 1–34). Springer US. doi: 10.1007/978-1-4899-7637-6_1

Richardson, R., Schultz, J., & Crawford, K. (2019). Dirty Data, Bad Predictions: How Civil Rights Violations Impact Police Data, Predictive Policing Systems, and Justice. New York University Law Review Online, Forthcoming.

Roff, H. M., & Moyes, R. (2016). Meaningful Human Control, Artificial Intelligence and Autonomous Weapons. In Proceedings of the 2016 International Committee of the Red Cross.

Romero, S. (2018). Self-Driving Uber Car Kills Pedestrian in Arizona, Where Robots Roam. The New York Times. https://www.nytimes.com/2018/03/19/technology/uber-driverless-fatality.html

Rometty, G. (2018). IBM CEO Ginni Rometty's Letter to the U.S. Congress. IBM Policy Lab.

ROSS, C., & SWETLITZ, I. (2017). *IBM PITCHED ITS WATSON SUPERCOMPUTER AS A REVOLUTION IN CANCER CARE. IT'S NOWHERE CLOSE.* STAT NEWS. HTTPS://WWW.STATNEWS.COM/2017/09/05/WATSON-IBM-CANCER/

RUDER, S., GHAFFARI, P., & BRESLIN, J. G. (2016). *A HIERARCHICAL MODEL OF REVIEWS FOR ASPECT-BASED SENTIMENT ANALYSIS. EMNLP 2016 - 2016 CONFERENCE ON EMPIRICAL METHODS IN NATURAL LANGUAGE PROCESSING, PROCEEDINGS,* 999–1005.

RUDIN, C. (2019). *STOP EXPLAINING BLACK BOX MACHINE LEARNING MODELS FOR HIGH STAKES DECISIONS AND USE INTERPRETABLE MODELS INSTEAD. NATURE MACHINE INTELLIGENCE,* 1(5), 206-215.

RUSSELL, S. (2019). *HUMAN COMPATIBLE: ARTIFICIAL INTELLIGENCE AND THE PROBLEM OF CONTROL.* VIKING.

RUSSELL, S. J., & NORVIG, P. (2016). *"ARTIFICIAL INTELLIGENCE: A MODERN APPROACH."* PEARSON.

RUSSELL, S. J., & NORVIG, P. (2016). *ARTIFICIAL INTELLIGENCE: A MODERN APPROACH.* PEARSON.

RUSSELL, S. J., & NORVIG, P. (2016). *ARTIFICIAL INTELLIGENCE: A MODERN APPROACH.* MALAYSIA; PEARSON EDUCATION LIMITED.

RUSSELL, S. J., DEWEY, D., & TEGMARK, M. (2015). *RESEARCH PRIORITIES FOR ROBUST AND BENEFICIAL ARTIFICIAL INTELLIGENCE. AI MAGAZINE,* 36(4), 105-114.A

RUSSELL, S. J., NORVIG, P., DAVIS, E., & OTHERS. (2010). *ARTIFICIAL INTELLIGENCE: A MODERN APPROACH.* PEARSON.

RUSSELL, S., DEWEY, D., & TEGMARK, M. (2015). *RESEARCH PRIORITIES FOR ROBUST AND BENEFICIAL ARTIFICIAL INTELLIGENCE. AI MAGAZINE,* 36(4), 105-114.

SAHAY, R. (2019). *FIGHTING FRAUD IN THE E-COMMERCE CHANNEL: A MERCHANT'S GUIDE.* BUSINESS EXPERT PRESS.

SAMUEL, A. L. (1959). *"SOME STUDIES IN MACHINE LEARNING USING THE GAME OF CHECKERS." IBM JOURNAL OF RESEARCH AND DEVELOPMENT,* 3(3), 210-229.

SCHAAR, P. (2010). *PRIVACY BY DESIGN. IDENTITY IN THE INFORMATION SOCIETY,* 3(2), 267–274. HTTPS://DOI.ORG/10.1007/S12394-010-0062-Y

SCHMIDHUBER, J. (2015). *DEEP LEARNING IN NEURAL NETWORKS: AN OVERVIEW. NEURAL NETWORKS,* 61, 85-117.

SCHOLEM, G. (1991). *ON THE KABBALAH AND ITS SYMBOLISM.* SCHOCKEN BOOKS.

SCHREIER, J. (2018). THE MAKING OF ROCKSTAR GAMES' RED DEAD REDEMPTION 2. KOTAKU. RETRIEVED FROM: HTTPS://KOTAKU.COM/THE-MAKING-OF-ROCKSTAR-GAMES-RED-DEAD-REDEMPTION-2-1838778228

SCHROEPFER, M. (2020). AN UPDATE ON OUR WORK TO KEEP PEOPLE SAFE AND MAINTAIN INTEGRITY ON FACEBOOK. FACEBOOK NEWSROOM.

SCHWARTZ, P. M. (2019). DATA PROTECTION LAW AND THE ETHICAL USE OF DATA. BERKELEY TECHNOLOGY LAW JOURNAL, 34(2), 287-296.

SCHWARTZ, P. M., & JAHN, F. (2020). TRANSATLANTIC DATA PRIVACY LAW. GEORGETOWN LAW JOURNAL, 106, 115-179.

SENIOR, A. W., EVANS, R., JUMPER, J., KIRKPATRICK, J., SIFRE, L., GREEN, T., ... & HASSABIS, D. (2020). IMPROVED PROTEIN STRUCTURE PREDICTION USING POTENTIALS FROM DEEP LEARNING. NATURE, 577(7792), 706-710.

SHALLUE, C. J., & VANDERBURG, A. (2018). IDENTIFYING EXOPLANETS WITH DEEP LEARNING: A FIVE-PLANET RESONANT CHAIN AROUND KEPLER-80 AND AN EIGHTH PLANET AROUND KEPLER-90. THE ASTRONOMICAL JOURNAL, 155(2), 94. HTTPS://DOI.ORG/10.3847/1538-3881/AA9E09

SHEPARDSON, D. (2021). U.S. OPENS SAFETY PROBE INTO TESLA AUTOPILOT CRASHES WITH EMERGENCY VEHICLES. REUTERS. HTTPS://WWW.REUTERS.COM/BUSINESS/AUTOS-TRANSPORTATION/US-OPEN-PROBE-INTO-11-TESLA-CRASHES-SINCE-2018-INVOLVING-EMERGENCY-VEHICLES-2021-08-16/

SIEGWART, R., & NOURBAKHSH, I. R. (2004). INTRODUCTION TO AUTONOMOUS ROBOTS: KINEMATICS, PERCEPTION, LOCALIZATION AND PLANNING. MIT PRESS.

SIEMENS. (2021). HOW ARTIFICIAL INTELLIGENCE CAN BOOST YOUR PRODUCTION. SIEMENS DIGITAL INDUSTRIES. HTTPS://NEW.SIEMENS.COM/GLOBAL/EN/PRODUCTS/AUTOMATION/TECHNICAL/AI-IN-MANUFACTURING.HTML

SILVER, D., HUANG, A., MADDISON, C. J., GUEZ, A., SIFRE, L., VAN DEN DRIESSCHE, G., ... & DIELEMAN, S. (2016). MASTERING THE GAME OF GO WITH DEEP NEURAL NETWORKS AND TREE SEARCH. NATURE, 529(7587), 484-489.

SIMON, H. A. (1965). THE SHAPE OF AUTOMATION FOR MEN AND MANAGEMENT. HARPER & ROW.

SMITH, A. (2019). ARTIFICIAL INTELLIGENCE: THE EVOLUTION OF SIRI, ALEXA, AND GOOGLE ASSISTANT. FORBES.

SMITH, B. (2018). ADVANCING OUR APPROACH TO AI. MICROSOFT ON THE ISSUES.

SMITH, B. (2019). *TOOLS AND WEAPONS: THE PROMISE AND THE PERIL OF THE DIGITAL AGE. PENGUIN BOOKS.*

SMITH, B., & BROWNE, C. (2019). *TOOLS AND WEAPONS: THE PROMISE AND THE PERIL OF THE DIGITAL AGE. PENGUIN BOOKS.*

SMITH, C. (2016). *AMAZON'S RECOMMENDATION SECRET. BBC NEWS.* HTTPS://WWW.BBC.COM/NEWS/BUSINESS-35360375

SMITH, M. D., BAILEY, J., & BRYNJOLFSSON, E. (2017). *UNDERSTANDING DIGITAL MARKETS: REVIEW AND ASSESSMENT. IN ECONOMICS OF THE INFORMATION INDUSTRY.*

STONE, M. (2020). *AI IS HELPING TURN THE TIDE AGAINST FOOD WASTE IN FARMING. SILICON ANGLE.* HTTPS://SILICONANGLE.COM/2020/09/28/AI-HELPING-TURN-TIDE-FOOD-WASTE-FARMING/

STOYCHEFF, E., LIU, J., & WIBOWO, K. (2020). *UNDER SURVEILLANCE: EXAMINING FACEBOOK'S SPIRAL OF SILENCE EFFECTS IN THE WAKE OF NSA INTERNET MONITORING. JOURNALISM & MASS COMMUNICATION QUARTERLY, 93(1), 16–37.*

SULEYMAN, M., AND HASSABIS, D. *"DEEPMIND AI REDUCES GOOGLE DATA CENTRE COOLING BILL BY 40%." DEEPMIND BLOG. (2016).*

SUTTON, R. S., & BARTO, A. G. (2018). *REINFORCEMENT LEARNING: AN INTRODUCTION. MIT PRESS.*

SZELISKI, R. (2010). *COMPUTER VISION: ALGORITHMS AND APPLICATIONS. SPRINGER.*

TADDEO, M., & FLORIDI, L. (2018). *REGULATE ARTIFICIAL INTELLIGENCE TO AVERT CYBER ARMS RACE. NATURE, 556(7701), 296-298.*

TEGMARK, M. (2017). *LIFE 3.0: BEING HUMAN IN THE AGE OF ARTIFICIAL INTELLIGENCE. KNOPF.*

TESLA (2021). *AUTOPILOT AND FULL SELF-DRIVING CAPABILITY.* HTTPS://WWW.TESLA.COM/AUTOPILOT

TRICOG (2020). *HOW TRICOG WORKS.* HTTPS://TRICOG.COM/HOW-TRICOG-WORKS/

TURING, A. M. (1936). *ON COMPUTABLE NUMBERS, WITH AN APPLICATION TO THE ENTSCHEIDUNGSPROBLEM. PROCEEDINGS OF THE LONDON MATHEMATICAL SOCIETY, 42(2), 230-265.*

TURING, A. M. (1950). *"COMPUTING MACHINERY AND INTELLIGENCE." MIND, 59(236), 433-460.*

TURING, A. M. (1950). *COMPUTING MACHINERY AND INTELLIGENCE. MIND, 59(236), 433-460.*

TURNER, A. (2011). GREEK AUTOMATA: AN ANCIENT TECH REVOLUTION. THE GUARDIAN.

UDEMY. WWW.UDEMY.COM

UJUZIKILIMO (2021). UJUZIKILIMO SOLUTIONS. HTTPS://WWW.UJUZIKILIMO.COM/

UK DEPARTMENT FOR EDUCATION. (2019). REALIZING THE POTENTIAL OF TECHNOLOGY IN EDUCATION: A STRATEGY FOR EDUCATION PROVIDERS AND THE TECHNOLOGY INDUSTRY.

UNESCO. (2020). HOW MUCH DOES YOUR COUNTRY INVEST IN R&D? UNESCO. HTTPS://DATA.UIS.UNESCO.ORG/INDEX.ASPX?DATASETCODE=SCN_DS&LANG=EN

UNITED STATES NAVY. (2021). AEGIS COMBAT SYSTEM. RETRIEVED FROM HTTPS://WWW.NAVY.MIL/RESOURCES/FACT-FILES/DISPLAY-FACTFILES/ARTICLE/2167793/AEGIS-COMBAT-SYSTEM/

UNIVERSITY OF CALIFORNIA, BERKELEY. (N.D.). ETHICS IN AI. RETRIEVED FROM HTTPS://WWW.BERKELEY.EDU/DEGREE/DATA-SCIENCE/MASTERS-PROGRAM/ETHICS-IN-AI

VAKIL, R. (2018). THE IMPORTANCE OF CODING IN THE WORKFORCE. SKILLCRUSH. RETRIEVED FROM HTTPS://SKILLCRUSH.COM/BLOG/IMPORTANCE-OF-CODING-IN-THE-WORKFORCE/

VAKIL, S. (2018). CODING FOR ALL: A DISTRICT-WIDE STUDY OF CODING INSTRUCTION. PROCEEDINGS OF THE 49TH ACM TECHNICAL SYMPOSIUM ON COMPUTER SCIENCE EDUCATION, 665-670. HTTPS://DOI.ORG/10.1145/3159450.3159608

VALVE. (2008). THE AI DIRECTOR IN LEFT 4 DEAD. VALVE DEVELOPER COMMUNITY. RETRIEVED FROM: HTTPS://DEVELOPER.VALVESOFTWARE.COM/WIKI/THE_AI_DIRECTOR_IN_LEFT_4_DEAD

VAN BUITENEN, J. A. B. (1973). THE MAHABHARATA: VOLUME 1. UNIVERSITY OF CHICAGO PRESS.

VAN DEN OORD, A., DIELEMAN, S., & SCHRAUWEN, B. (2013). DEEP CONTENT-BASED MUSIC RECOMMENDATION. IN PROCEEDINGS OF THE 26TH INTERNATIONAL CONFERENCE ON NEURAL INFORMATION PROCESSING SYSTEMS - VOLUME 2 (PP. 2643–2651). CURRAN ASSOCIATES INC.

VARIAN, H. R. (2014). BIG DATA: NEW TRICKS FOR ECONOMETRICS. JOURNAL OF ECONOMIC PERSPECTIVES, 28(2), 3-28.

VELOSO, M. (2014). "COGNITIVE COMPUTING." COMMUNICATIONS OF THE ACM, 57(10), 78-86.

VIHAVAINEN, A., LUUKKAINEN, M., KURHILA, J., & PAKSULA, M. (2013). MULTI-FACETED SUPPORTED ONLINE LEARNING: LESSONS LEARNED FROM INTRODUCTORY PROGRAMMING. IN PROCEEDINGS OF THE 13TH KOLI CALLING INTERNATIONAL CONFERENCE ON COMPUTING EDUCATION RESEARCH (PP. 89-98). ACM.

VIHAVAINEN, A., PAKSULA, M., & LUUKKAINEN, M. (2013). MOOC AS A REMEDY FOR THE SHORTAGE OF COMPUTER SCIENCE TEACHERS. ACM TRANSACTIONS ON COMPUTING EDUCATION (TOCE), 13(4), 1-18. HTTPS://DOI.ORG/10.1145/2534965

VINCENT, J. (2016). MICROSOFT'S AI BOT TAY GENERATES CONTROVERSY WITH RACIST TWEETS. THE VERGE. RETRIEVED FROM HTTPS://WWW.THEVERGE.COM/2016/3/24/11297050/TAY-MICROSOFT-CHATBOT-RACIST

VINCENT, J. (2016). TWITTER TAUGHT MICROSOFT'S AI CHATBOT TO BE A RACIST ASSHOLE IN LESS THAN A DAY. THE VERGE. HTTPS://WWW.THEVERGE.COM/2016/3/24/11297050/TAY-MICROSOFT-CHATBOT-RACIST

VINYALS, O., BABUSCHKIN, I., CZARNECKI, W. M., MATHIEU, M., DUDZIK, A., CHUNG, J., ... & POWELL, R. (2019). GRANDMASTER LEVEL IN STARCRAFT II USING MULTI-AGENT REINFORCEMENT LEARNING. NATURE, 575(7782), 350-354.

WALKINGTON, C. (2013). USING ADAPTIVE LEARNING TECHNOLOGIES TO PERSONALIZE INSTRUCTION TO STUDENT INTERESTS: THE IMPACT OF RELEVANT CONTEXTS ON PERFORMANCE AND LEARNING OUTCOMES. JOURNAL OF EDUCATIONAL PSYCHOLOGY, 105(4), 932-945.

WANG, Y., CHEN, X., SKIENA, S., & HOOS, H. H. (2019). GOOGLE'S DEEPMIND AND THE FUTURE OF ARTIFICIAL INTELLIGENCE. IN PROCEEDINGS OF THE 24TH ACM SIGKDD INTERNATIONAL CONFERENCE ON KNOWLEDGE DISCOVERY & DATA MINING - KDD '18 (PP. 2435–2435). ACM PRESS. DOI: 10.1145/3219819.3226074

WANG, Y., ET AL. (2019). AI EDUCATION IN CHINA: CHALLENGES AND PERSPECTIVES. CHINA AI DEVELOPMENT REPORT.

WELLER, A., DOŠILOVIĆ, F., & BAUER, A. (2019). CHALLENGES FOR TRANSPARENCY. IN PROCEEDINGS OF THE 2019 AAAI/ACM CONFERENCE ON AI, ETHICS, AND SOCIETY (PP. 291–297).

WELLER, M. (2020). 25 YEARS OF ED TECH. ATHABASCA UNIVERSITY PRESS. HTTPS://WWW.AUPRESS.CA/BOOKS/120290-25-YEARS-OF-ED-TECH/

WINFIELD, L., & BISHOP, G. (2020). DEEP LEARNING FOR VISUAL EFFECTS. IN SIGGRAPH ASIA 2020 TECHNICAL BRIEFS (PP. 1–4). ACM. DOI: 10.1145/3415264.3442030

WITTEN, I. H., FRANK, E., HALL, M. A., & PAL, C. J. (2016). DATA MINING: PRACTICAL MACHINE LEARNING TOOLS AND TECHNIQUES (4TH ED.). MORGAN KAUFMANN.

WORLD BANK. (2021). ACCESS TO ELECTRICITY (% OF POPULATION). WORLD BANK. HTTPS://DATA.WORLDBANK.ORG/INDICATOR/EG.ELC.ACCS.ZS

WORLD ECONOMIC FORUM (2020). THE FUTURE OF JOBS REPORT 2020. HTTP://WWW3.WEFORUM.ORG/DOCS/WEF_FUTURE_OF_JOBS_2020.PDF

WORLD ECONOMIC FORUM. (2020). JOBS OF TOMORROW: MAPPING OPPORTUNITY IN THE NEW ECONOMY. WORLD ECONOMIC FORUM. HTTPS://WWW.WEFORUM.ORG/REPORTS/JOBS-OF-TOMORROW-MAPPING-OPPORTUNITY-IN-THE-NEW-ECONOMY

XIONG, W., DROPPO, J., HUANG, X., SEIDE, F., SELTZER, M., STOLCKE, A., ... & ZWEIG, G. (2016). ACHIEVING HUMAN PARITY IN CONVERSATIONAL SPEECH RECOGNITION. ARXIV PREPRINT ARXIV:1610.05256.

XU, A., LIU, Z., GUO, Y., SINHA, V., & AKKIRAJU, R. (2021). A NEW CHATBOT FOR CUSTOMER SERVICE ON SOCIAL MEDIA. IN PROCEEDINGS OF THE 2017 CHI CONFERENCE ON HUMAN FACTORS IN COMPUTING SYSTEMS (PP. 3506-3510).

ZEMEL, R., WU, Y., SWERSKY, K., PITASSI, T., & DWORK, C. (2013). LEARNING FAIR REPRESENTATIONS. IN PROCEEDINGS OF THE 30TH INTERNATIONAL CONFERENCE ON MACHINE LEARNING (ICML) (VOL. 28, NO. 3, PP. 325-333).

ZHANG, X., KIM, J., PATZER, R. E., PITTS, S. R., PATZER, A., & SCHRAGER, J. D. (2020). PREDICTION OF EMERGENCY DEPARTMENT HOSPITAL ADMISSION BASED ON NATURAL LANGUAGE PROCESSING AND NEURAL NETWORKS. METHODS OF INFORMATION IN MEDICINE, 57(05/06), 141-147.

ZHAVORONKOV, A., IVANENKOV, Y. A., ALIPER, A., VESELOV, M. S., ALADINSKIY, V. A., ALADINSKAYA, A. V., ... & ZHOLUS, A. (2019). DEEP LEARNING ENABLES RAPID IDENTIFICATION OF POTENT DDR1 KINASE INHIBITORS. NATURE BIOTECHNOLOGY, 37(9), 1038-1040.

ZHOU, C., SU, F., PEI, T., ZHANG, A., DU, Y., LUO, B., CAO, Z., WANG, J., YUAN, W., ZHU, Y., SONG, C., CHEN, J., XU, J., LI, F., MA, T., JIANG, L., YAN, F., YI, J., HU, Y., & LIAO, Y. (2020). COVID-19: CHALLENGES TO GIS WITH BIG DATA. GEOGRAPHY AND SUSTAINABILITY, 1(1), 77-87.

ZHOU, Y., & JIANG, X. (2012). DISSECTING ANDROID MALWARE: CHARACTERIZATION AND EVOLUTION. 2012 IEEE SYMPOSIUM ON SECURITY AND PRIVACY, 95–109. DOI: 10.1109/SP.2012.16

ZIPLINE (2020). HOW IT WORKS. HTTPS://FLYZIPLINE.COM/HOW-IT-WORKS/

ZIPLINE (2021). ZIPLINE'S IMPACT IN GHANA. HTTPS://FLYZIPLINE.COM/GHANA/

ZUBOFF, S. (2019). THE AGE OF SURVEILLANCE CAPITALISM: THE FIGHT FOR A HUMAN FUTURE AT THE NEW FRONTIER OF POWER. PUBLIC AFFAIRS.

Appendices

Glossary of AI Terms

Active Learning: A machine learning approach that involves an AI system actively selecting and acquiring labeled data from a human expert or oracle to improve its performance. Active learning reduces the labeling burden and enhances the learning process.

Adaptability: The ability of AI systems or algorithms to learn and adjust their behavior based on new data or changing circumstances. Adaptive AI systems can improve performance over time and adjust to dynamic environments.

Adversarial Machine Learning: A field of study within AI that focuses on understanding and defending against adversarial attacks. Adversarial machine learning aims to develop robust models that can withstand malicious attempts to manipulate or deceive the AI system.

Aeolipile, also known as a Hero's engine or Hero's turbine, is a simple ancient device that demonstrates the principle of jet propulsion. It was invented by the Greek mathematician and engineer Hero of Alexandria in the first century AD.

Agent: An autonomous entity or program that acts on behalf of a user or system in AI. Agents can perceive their environment, make decisions, and take actions to achieve specific goals.

AI Ethics: The branch of ethics that focuses on the ethical implications, considerations, and guidelines related to the development, deployment, and use of AI systems. AI ethics addresses concerns such as transparency, fairness, privacy, accountability, and bias.

Algorithm: A step-by-step procedure or set of rules for solving a problem or performing a specific task. In AI, algorithms are designed to enable machines to process data, make decisions, and learn from patterns.

Ambient Intelligence: The concept of creating intelligent and responsive environments that can perceive, understand, and adapt to human needs. Ambient intelligence aims to integrate AI and technology seamlessly into the surroundings, enhancing human experiences and interactions.

Anomaly Detection: The process of identifying unusual or abnormal patterns or data points within a dataset. AI algorithms can learn from normal patterns and detect deviations that may indicate potential fraud, errors, or outliers.

Argumentation: A field of AI that deals with the representation and analysis of arguments. Argumentation models are used to evaluate and reason with conflicting viewpoints or claims, enabling AI systems to engage in logical debates or decision-making processes.

Artificial Consciousness: The theoretical concept of creating or simulating consciousness in an AI system. Artificial consciousness explores the idea of self-awareness, subjective experience, and the ability to perceive and understand the world.

Artificial General Intelligence (AGI): An AI system or entity that possesses general intelligence comparable to human intelligence. AGI can understand, learn, and apply knowledge across a wide range of tasks and domains.

Artificial Intelligence: The field of computer science that focuses on creating intelligent machines capable of performing tasks that typically require human intelligence, such as reasoning, learning, problem-solving, and decision-making.

Artificial Neural Architecture Search (ANAS): The process of using AI algorithms to automatically search and design the optimal architecture or structure of artificial neural networks. ANAS aims to improve network performance and efficiency.

Artificial Neural Network (ANN): A computational model inspired by the structure and function of biological neural networks. ANNs are used in machine learning and deep learning to recognize patterns, make predictions, and solve complex problems.

Association Rule Mining: A data mining technique in AI that discovers interesting relationships, patterns, or associations among variables or items within a dataset. It is often used for market basket analysis or recommendation systems.

Augmented Reality (AR): A technology that combines the real world with virtual elements, allowing computer-generated objects or information to be overlaid on the physical environment. AR can enhance human perception and interaction with the environment, often using AI algorithms.

Autoencoder: A type of neural network architecture used for unsupervised learning. Autoencoders aim to learn efficient representations or compressed versions of input data by reconstructing the input from a lower-dimensional latent space.

Automata is an automaton, a machine or a mathematical model that can perform tasks or exhibit behavior without the need for constant human intervention. It is designed to follow a set of predetermined rules or instructions.

Automated Reasoning: The process of using AI and automated techniques to perform logical reasoning and draw conclusions from given facts or knowledge. Automated reasoning is used in various applications, including theorem proving and expert systems.

Automation: The process of using AI and technology to automate tasks and processes that were previously performed by humans. It involves the use of software, robots, or other AI systems to perform repetitive or complex tasks.

Backpropagation: A training algorithm used in artificial neural networks to adjust the weights and biases based on the error between the predicted output and the desired output. Backpropagation is crucial for learning and improving the performance of neural networks.

Bagging: An ensemble learning technique in machine learning where multiple models are trained independently on different subsets of the training data. The predictions from the individual models are combined to make a final prediction, improving overall performance.

Bayesian Inference: A statistical method used in AI and machine learning to update probabilities and beliefs based on new evidence or data. Bayesian inference allows for incorporating prior knowledge and adjusting beliefs through a probabilistic framework.

Bayesian Network: A probabilistic graphical model that represents the relationships between variables using a directed acyclic graph. Bayesian networks are used for reasoning under uncertainty and making predictions based on observed evidence.

Bayesian Optimization: A technique that uses Bayesian methods to optimize and search for the best configuration or set of parameters for a given function or algorithm. Bayesian optimization is commonly used in hyperparameter tuning for machine learning models.

Behavior Analysis: The process of analyzing and interpreting patterns, behaviors, and interactions of individuals or systems. AI techniques, such as machine learning and computer vision, are applied to behavior analysis for applications like surveillance and anomaly detection.

Behavior Cloning: A technique in machine learning where an AI system learns to imitate or replicate human behavior by training on labeled examples provided by human experts. Behavior cloning is commonly used in autonomous driving and robotics.

Behavior Trees: A hierarchical model used to represent and control the behavior of autonomous agents or AI systems. Behavior trees define a set of tasks and conditions, enabling the AI system to make decisions and take actions based on a predefined structure.

Bias: In AI, bias refers to the systematic and unfair preferences or prejudices that an AI system may exhibit due to the data it was trained on or the design choices made. Addressing bias is crucial for developing ethical and fair AI systems.

Big Data: Refers to extremely large and complex datasets that are difficult to process and analyze using traditional methods. AI techniques, such as machine learning and data mining, are often employed to extract valuable insights and patterns from big data.

Bioinformatics: The application of AI and computational methods to analyze biological data, such as genomic sequences and protein structures. Bioinformatics plays a vital role in genomics research, drug discovery, and personalized medicine.

Biometric Identification: The use of AI and machine learning algorithms to identify and authenticate individuals based on their unique physiological or behavioral characteristics, such as fingerprints, facial features, or voice patterns.

Biomimicry: The practice of emulating or drawing inspiration from nature and biological systems to design and develop AI algorithms or systems. Biomimicry explores natural processes, behaviors, and structures to enhance AI capabilities.

Black Box: Refers to an AI system or model that operates as a closed system, where the internal workings or decision-making processes are not transparent or easily explainable. Understanding and interpreting black box models is an active area of research in AI.

Blockchain: A decentralized and distributed digital ledger technology that securely records and verifies transactions across multiple computers or nodes. AI and blockchain integration have the potential to enhance security, privacy, and trust in various applications.

Boolean Algebra is a mathematical system that deals with logical statements or variables that can have only two possible values: true or false, represented as 1 or 0. It provides a set of rules and operations to manipulate and analyze these logical values. In simple terms, Boolean algebra helps us reason and solve problems using simple rules of logic, such as AND, OR, and NOT, which determine how true and false values can be combined or negated. It forms the foundation of digital electronics and computer science, where binary logic is used to represent and process information.

Bootstrapping: A method in AI that involves starting with a small amount of labeled data and gradually expanding the training dataset through iterative processes. Bootstrapping is useful when labeled data is limited or expensive to obtain.

Bot: A software application or AI agent that performs automated tasks or interacts with users through conversational interfaces. Bots can be found in chatbots, virtual assistants, customer service systems, and social media platforms.

Brain-Computer Interface (BCI): A system that allows direct communication or interaction between the brain and external devices using AI algorithms and neurophysiological signals. BCI holds promise for applications like prosthetics control and assistive technologies.

Business Intelligence: The use of AI, data analytics, and visualization techniques to analyze and extract insights from large volumes of business data. Business intelligence helps organizations make informed decisions, identify trends, and improve operational efficiency.

Chatbot: A computer program or AI agent designed to simulate human conversation and interact with users through text or voice-based interfaces. Chatbots use natural language processing and machine learning techniques to understand and respond to user queries.

Cloud Computing: The practice of using remote servers and networks, often accessed through the internet, to store, process, and analyze data. Cloud computing provides on-demand access to computing resources, enabling scalable AI applications.

Clustering: A technique in machine learning that groups similar data points together based on their characteristics or similarities. Clustering algorithms help discover patterns and structures within datasets without predefined class labels.

Cognitive Computing: A branch of AI that aims to create systems that can mimic human cognitive processes, including perception, reasoning, learning, and problem-solving. Cognitive computing systems leverage machine learning, natural language processing, and other AI techniques.

Collaborative Filtering: A technique in recommendation systems that predicts user preferences or recommendations based on similarities and patterns observed from the preferences of other users. Collaborative filtering is commonly used in personalized recommendation engines.

Computer Vision: The field of AI that focuses on enabling computers or machines to understand, analyze, and interpret visual information from images or videos. Computer vision involves tasks such as object recognition, image segmentation, and scene understanding.

Constraint Satisfaction: The process of finding solutions or assignments to a set of variables that satisfy a given set of constraints. Constraint satisfaction problems arise in various AI domains, such as scheduling, planning, and optimization.

Convolutional Neural Network (CNN): A type of deep learning neural network specifically designed for processing and analyzing grid-like data, such as images or video frames. CNNs use convolutional layers to extract relevant features from input data.

Cybersecurity: The practice of protecting computer systems, networks, and data from unauthorized access, attacks, or breaches. AI is increasingly employed in cybersecurity to detect and respond to threats, identify anomalies, and enhance security measures.

Data Augmentation: A technique used in machine learning to artificially increase the size or diversity of a training dataset by creating variations of the existing data. Data augmentation helps improve model generalization and robustness.

Data Governance: The framework, policies, and processes for managing and ensuring the quality, integrity, and security of data used in AI systems. Data governance helps maintain data reliability and compliance with regulations.

Data Labeling: The process of annotating or assigning labels to data instances for supervised learning. Data labeling is crucial for creating labeled datasets that can be used to train AI models.

Data Mining: The process of discovering patterns, relationships, or insights from large volumes of structured or unstructured data. Data mining techniques, such as clustering, classification, and association rule mining, are used to extract knowledge from data.

Data Preprocessing: The process of cleaning, transforming, and organizing raw data before it is used for analysis or training AI models. Data preprocessing includes tasks such as data cleaning, feature scaling, and handling missing values.

Data Synthesis: The process of generating synthetic data that closely resembles real-world data. Data synthesis techniques are used when access to real data is limited, and AI models need additional data for training or evaluation.

Datasets: Collections of data that are used for training, testing, or validating AI models. Datasets can include various types of data, such as images, text, audio, or video, and are essential for developing and evaluating AI systems.

Decision Tree: A machine learning algorithm that builds a tree-like model of decisions and their possible consequences based on input features. Decision trees are used for classification, regression, and decision-making tasks in AI.

Deep Learning: A subfield of machine learning that focuses on training artificial neural networks with multiple layers to learn and represent complex patterns and hierarchies in data. Deep learning has achieved remarkable success in various AI applications.

Deep Q-Network (DQN): A reinforcement learning algorithm that combines deep neural networks with Q-learning. DQNs are used to train agents that can make decisions and take actions in complex environments.

Deep Reinforcement Learning: A combination of deep learning and reinforcement learning techniques where an AI agent learns to make sequential decisions in an environment to maximize a reward signal. Deep reinforcement learning has achieved significant breakthroughs in areas like game playing and robotics.

Dempster-Shafer Theory: A mathematical framework used for reasoning under uncertainty and handling subjective or incomplete information. Dempster-Shafer theory is employed in AI for decision-making and belief updating.

Diagnostics: The process of identifying and diagnosing issues or problems in AI systems. Diagnostics involve analyzing system performance, error messages, and other indicators to detect and resolve issues.

Dialogue Systems: AI systems that enable natural language conversations between humans and machines. Dialogue systems, including chatbots and virtual assistants, use techniques from natural language processing and machine learning to understand and generate human-like responses.

Differential Privacy: A privacy-preserving technique in AI that aims to protect sensitive information in datasets. Differential privacy adds noise or perturbations to data to ensure that individual records cannot be distinguished.

Digital Twin: A virtual replica or simulation of a physical object, system, or process. Digital twins leverage IoT sensors, AI algorithms, and real-time data to enable monitoring, analysis, and optimization of physical entities.

Dimensionality Reduction: The process of reducing the number of input variables or features in a dataset. Dimensionality reduction techniques aim to eliminate irrelevant or redundant features while preserving the most important information.

DNN (Deep Neural Network): A neural network architecture with multiple hidden layers between the input and output layers. DNNs are capable of learning and representing intricate patterns and relationships in data.

Domain Adaptation: The process of transferring knowledge or models learned from one domain to another related domain. Domain adaptation helps in applying AI models trained on one set of data to new or different datasets.

Domain Knowledge: Expertise or specific knowledge about a particular field, industry, or domain. Domain knowledge is often incorporated into AI systems to improve their performance and accuracy in specific application areas.

Dynamic Programming: A technique in AI that solves complex problems by breaking them down into smaller overlapping subproblems. Dynamic programming involves storing and reusing solutions to subproblems to improve efficiency.

Early Stopping: A technique used in training machine learning models to prevent overfitting. Early stopping stops the training process when the model's performance on a validation set no longer improves, thus avoiding excessive training.

Edge AI: The deployment of AI algorithms and models on edge devices, such as smartphones, IoT devices, or edge servers, rather than relying solely on cloud-based processing. Edge AI enables faster, offline, and privacy-preserving AI applications.

Edge Computing: The practice of processing and analyzing data on edge devices, such as sensors or IoT devices, rather than sending it to a central server or cloud. Edge computing reduces latency and enhances privacy in AI applications.

Embedding: In natural language processing and machine learning, embedding refers to representing words, phrases, or entities as dense vectors in a continuous vector space. Embeddings capture semantic relationships and enable better analysis of textual data.

Emotion Recognition: The process of detecting and understanding human emotions, often through facial expressions, speech patterns, or physiological signals. Emotion recognition is used in AI applications like sentiment analysis and affective computing.

Ensemble Learning: A technique in machine learning that combines multiple models or classifiers to improve overall performance and accuracy. Ensemble learning methods include bagging, boosting, and stacking.

Ensemble Methods: Techniques that combine multiple models or predictions to improve the accuracy and robustness of AI systems. Ensemble methods include methods like bagging, boosting, and voting.

Entity Recognition: The task of identifying and classifying named entities, such as persons, organizations, locations, or dates, within text data. Entity recognition is a fundamental component of natural language processing and information extraction systems.

Episodic Memory: The type of memory that enables AI systems to store and recall specific events or episodes. Episodic memory is crucial for tasks requiring context and sequential decision-making.

Error Analysis: The process of identifying, analyzing, and understanding errors or mistakes made by AI systems. Error analysis helps in improving model performance, identifying biases, and refining algorithms.

Ethical AI: The field of AI that addresses the ethical implications, considerations, and guidelines for the development, deployment, and use of AI systems. Ethical AI focuses on fairness, transparency, accountability, privacy, and bias mitigation.

Event Detection: The task of identifying and classifying specific events or occurrences within a stream of data, such as detecting anomalies, trends, or critical incidents. Event detection is used in various AI applications, including surveillance and monitoring.

Event-Driven Architecture: An architectural pattern where software components or services communicate and respond to events or messages. Event-driven architectures are used in AI systems to handle asynchronous and real-time data processing.

Evolutionary Algorithms: Optimization algorithms inspired by the process of natural selection and evolution. Evolutionary algorithms use techniques like mutation, crossover, and selection to find optimal solutions to complex problems.

Evolutionary Computation: A subfield of AI that employs evolutionary algorithms and techniques, such as genetic algorithms, to solve complex optimization and search problems. Evolutionary computation is inspired by biological evolution and natural selection.

Experimentation: The process of conducting controlled experiments or tests to evaluate and compare different AI models, algorithms, or configurations. Experimentation helps researchers and practitioners assess the performance and effectiveness of AI systems.

Expert System: An AI system that emulates the decision-making and problem-solving capabilities of human experts in a specific domain. Expert systems use knowledge bases, inference engines, and rule-based reasoning to provide expert-level advice or solutions.

Explainable AI: The field of AI that focuses on developing models and algorithms that can provide understandable explanations for their decisions and actions. Explainable AI aims to enhance transparency, interpretability, and trust in AI systems.

Explainable Reinforcement Learning: The area of research that aims to provide explanations for the decisions and actions made by reinforcement learning agents. Explainable reinforcement learning enables better understanding and interpretability of RL algorithms.

Extractive Summarization: A technique in natural language processing where the most important information is extracted from a text document to create a concise summary. Extractive summarization methods select and assemble key sentences or phrases from the original text.

Face Recognition: The technology that uses AI algorithms to identify or verify individuals based on facial features. Face recognition systems analyze facial patterns and compare them against a database for identification purposes.

Fairness: The ethical principle in AI that ensures unbiased and equitable treatment of individuals or groups. Fairness aims to mitigate biases and discriminatory effects in AI algorithms and decision-making systems.

Fallback Strategy: A contingency plan in AI systems that specifies alternative actions or responses when the system encounters unknown or unhandled situations. Fallback strategies help AI systems gracefully handle unexpected scenarios.

False Positive: In binary classification tasks, a false positive refers to a prediction or classification error where the model incorrectly identifies a negative sample as positive. False positives can have significant consequences in AI applications like medical diagnosis or security systems.

Fault Diagnosis: The process of identifying, analyzing, and diagnosing faults or failures in AI systems. Fault diagnosis involves detecting anomalies, troubleshooting issues, and maintaining system reliability.

Feature Engineering: The process of creating new features or transforming existing features to improve the performance of machine learning models. Feature engineering involves domain knowledge and understanding the problem at hand.

Feature Extraction: The process of selecting or extracting relevant features or characteristics from raw data. Feature extraction is an important step in machine learning to represent data in a more compact and meaningful way.

Feature Selection: The process of identifying the most relevant subset of features from a larger set for building machine learning models. Feature selection helps reduce dimensionality and improve model simplicity and interpretability.

Federated Database: A distributed database system where data is stored across multiple locations or nodes while maintaining data consistency and integrity. Federated databases enable efficient and secure data access in distributed AI systems.

Federated Learning: A distributed machine learning approach where AI models are trained across multiple devices or edge nodes without sharing raw data. Federated learning enables privacy-preserving collaborative learning.

Feedback Loop: The continuous cycle of receiving feedback, analyzing it, and making adjustments or improvements based on the feedback received. Feedback loops are essential in AI systems to iteratively refine performance and enhance learning.

Forecasting: The task of using historical data and AI methods to predict future trends or values. Forecasting techniques are widely used in time series analysis, sales predictions, demand forecasting, and resource planning.

Formal Verification: The process of rigorously proving or verifying the correctness of AI systems or algorithms using mathematical or logical methods. Formal verification ensures the absence of errors, bugs, or vulnerabilities.

Forward Chaining: A reasoning method where AI systems use available facts and rules to derive new conclusions or make inferences. Forward chaining starts with initial data and progresses forward to reach a conclusion.

Framework: A software development platform or infrastructure that provides libraries, tools, and components for building AI applications. Frameworks, such as TensorFlow and PyTorch, offer functionalities for training and deploying AI models.

Fraud Detection: The application of AI techniques to identify fraudulent activities or behaviors within a system or dataset. Fraud detection systems use patterns, anomalies, or machine learning algorithms to detect fraudulent patterns.

Function Approximation: The process of estimating or approximating a target function using a set of training examples or data points. Function approximation techniques, such as regression or neural networks, are used in supervised learning.

Functional Programming: A programming paradigm that emphasizes the use of pure functions and immutable data. Functional programming is relevant to AI development due to its focus on data transformations and composability.

Future Prediction: The task of using AI techniques to make predictions or forecasts about future events or outcomes based on historical data or patterns. Future prediction has applications in areas such as finance, weather forecasting, and stock market analysis.

Fuzzy Logic: A mathematical logic that deals with reasoning and decision-making in situations where uncertainty and imprecision are present. Fuzzy logic allows for the representation of partial truths and degrees of membership.

Game Theory: A mathematical framework used in AI to model and analyze strategic interactions between multiple agents or players. Game theory helps understand decision-making, competition, and cooperation in complex systems.

Gaussian Mixture Model (GMM): A probabilistic model that represents a probability distribution as a mixture of multiple Gaussian distributions. GMMs are used for tasks such as clustering and density estimation in machine learning.

Gaussian Process: A probabilistic model that defines a distribution over functions. Gaussian processes are used for regression, uncertainty estimation, and Bayesian optimization in machine learning and AI.

Generalization: The ability of an AI model to perform accurately on unseen or new data that was not present during training. Generalization is a key aspect of machine learning models to ensure their applicability to real-world scenarios.

Generative Adversarial Network (GAN): A type of deep learning model that consists of two neural networks, a generator and a discriminator, competing against each other.

GANs are used to generate synthetic data or images that resemble real data.

Genetic Algorithm: An optimization algorithm inspired by the process of natural selection and genetics. Genetic algorithms use techniques like mutation, crossover, and selection to find optimal solutions to complex problems.

Genetic Encoding: The representation of individuals or solutions in evolutionary algorithms using genetic codes, such as binary strings or tree structures. Genetic encoding helps define the search space and operators for evolving solutions.

Genetic Programming: A subfield of AI that applies evolutionary algorithms to automatically discover or evolve computer programs or algorithms. Genetic programming uses techniques like mutation and crossover to optimize program structures.

Gesture Recognition: The process of interpreting and understanding human gestures, typically performed by analyzing body movements or hand gestures. Gesture recognition is used in AI applications like human-computer interaction and sign language interpretation.

Gesture-Based Interaction: A form of human-computer interaction where users interact with computers or devices through gestures or body movements. Gesture-based interfaces are used in AI applications like virtual reality and augmented reality systems.

Goal-Oriented Dialogue Systems: AI systems that engage in natural language conversations with users to achieve specific goals or tasks. Goal-oriented dialogue systems are used in applications like virtual assistants and customer service chatbots.

Gradient Descent: An optimization algorithm used in training machine learning models to minimize the error or loss function. Gradient descent iteratively adjusts the model's parameters based on the negative gradient of the loss function.

Grammar Induction: The process of automatically inferring the grammatical rules or structures of a language from a set of observed sentences or text data. Grammar induction is a subfield of natural language processing.

Granger Causality: A statistical concept used to determine if one time series can predict or cause changes in another time series. Granger causality analysis helps identify causal relationships and dependencies in data.

Granular Computing: A methodology that deals with the representation, processing, and interpretation of complex information at different levels of granularity. Granular computing aims to handle uncertainty, vagueness, and imprecision in AI systems.

Graph Database: A database system that stores and represents data using a graph-based structure, consisting of nodes, edges, and properties. Graph databases are useful for modeling and querying complex relationships in AI applications.

Graph Neural Network (GNN): A type of neural network specifically designed for processing and learning from graph-structured data. GNNs operate on graphs and capture relational information between entities or nodes.

Greedy Algorithm: An algorithmic approach that makes locally optimal choices at each step in order to find an approximate solution. Greedy algorithms are often used for optimization problems in AI and can provide fast but not necessarily optimal solutions.

Grid Computing: A distributed computing infrastructure that allows the sharing and coordination of computational resources across multiple machines or nodes. Grid computing facilitates high-performance computing and data-intensive tasks in AI.

Haptic Feedback: The use of tactile or touch-based feedback to provide users with sensory information or cues in human-computer interaction. Haptic feedback enhances user experience and can be used in AI applications such as virtual reality and robotics.

Hardware Acceleration: The use of specialized hardware, such as GPUs (Graphics Processing Units) or TPUs (Tensor Processing Units), to accelerate the execution of AI computations. Hardware acceleration can significantly speed up AI training and inference processes.

Hebbian Learning: A learning rule in artificial neural networks that strengthens the connections between neurons when they are active simultaneously. Hebbian learning is based on the principle of synaptic plasticity and is used for unsupervised learning.

Heterogeneous Computing: The use of different types of computing devices or architectures, such as CPUs, GPUs, and specialized accelerators, to collectively solve computational problems. Heterogeneous computing can enhance AI performance and efficiency.

Heuristic: A problem-solving approach or rule of thumb that guides AI algorithms in making decisions or finding solutions. Heuristics provide practical strategies that may not guarantee optimal solutions but are often efficient.

Hierarchical Clustering: A clustering technique that groups similar data points into nested clusters or a hierarchy of clusters. Hierarchical clustering enables the identification of relationships and structures in datasets.

Hierarchical Reinforcement Learning: A reinforcement learning approach that involves learning and planning at multiple levels of abstraction. Hierarchical reinforcement learning enables agents to solve complex tasks by decomposing them into subtasks.

High-Dimensional Data: Data that has a large number of features or dimensions, often exceeding the number of samples or observations. High-dimensional data poses challenges in AI, requiring dimensionality reduction techniques and specialized algorithms.

High-Level Reasoning: The cognitive ability of AI systems to perform abstract and complex reasoning tasks, such as logical deduction, planning, and problem-solving. High-level reasoning is crucial for advanced AI capabilities, including decision-making and creativity.

HMM (Hidden Markov Model): A statistical model that represents a system with hidden states and observable outputs. HMMs are used in AI for tasks such as speech recognition, natural language processing, and gesture recognition.

Holistic Analysis: An approach in AI that considers the complete system or problem as a whole rather than focusing on isolated components. Holistic analysis involves understanding the interdependencies and interactions within the system for comprehensive understanding and decision-making.

Homomorphic Encryption: A cryptographic technique that allows computations to be performed on encrypted data without decrypting it. Homomorphic encryption preserves data privacy and security while enabling AI analysis on sensitive information.

Human Activity Recognition: The task of automatically recognizing and classifying human activities or actions from sensor data, such as accelerometer or video data. Human activity recognition is used in AI applications like healthcare, sports analysis, and surveillance.

Human Pose Estimation: The task of estimating and inferring the pose or body configuration of humans from images or video data. Human pose estimation enables applications such as motion capture, gesture recognition, and action recognition.

Human-AI Collaboration: The synergy between human intelligence and AI systems to jointly solve problems, make decisions, or perform tasks. Human-AI collaboration aims to leverage the complementary strengths of humans and machines for improved outcomes.

Human-AI Interaction: The study and design of interfaces and interactions between humans and AI systems. Human-AI interaction aims to create intuitive and effective ways for humans to communicate, collaborate, and control AI technologies.

Human-Centered AI: The design and development of AI systems with a focus on human needs, values, and ethical considerations. Human-centered AI aims to ensure that AI technologies align with human values and serve human interests.

Human-in-the-Loop (HITL): An AI system design where human intelligence and expertise are integrated into the decision-making loop. Human-in-the-loop approaches combine human judgment with AI algorithms to achieve more accurate and reliable outcomes.

Human-Level AI: The concept of AI systems achieving intelligence and cognitive capabilities comparable to human intelligence. Human-level AI aims to develop AI systems that exhibit advanced reasoning, learning, perception, and creativity similar to human capabilities.

Humanoid Robot: A robot designed to resemble and interact with humans in a human-like manner. Humanoid robots incorporate AI algorithms for perception, movement, and natural language processing to facilitate human-robot interaction.

Human-Robot Collaboration: The collaboration and interaction between humans and robots in shared workspaces or tasks. Human-robot collaboration involves AI techniques for safe, efficient, and intuitive cooperation between humans and robotic systems.

Hybrid Intelligence: The combination of human and artificial intelligence to create synergistic systems that leverage the strengths of both. Hybrid intelligence seeks to integrate human expertise and intuition with AI algorithms for enhanced decision-making and problem-solving.

Hybrid Model: A machine learning model that combines multiple algorithms, techniques, or modalities to improve overall performance. Hybrid models often integrate different AI approaches, such as combining neural networks with decision trees.

Hybrid Recommender System: A recommendation system that combines multiple approaches or techniques, such as collaborative filtering and content-based filtering, to provide personalized recommendations. Hybrid recommender systems leverage the strengths of different recommendation algorithms.

Hybridization: The process of combining different AI techniques, algorithms, or approaches to create hybrid models or systems that leverage the strengths of each component. Hybridization aims to achieve improved performance and versatility in AI applications.

Hypergraph Learning: A branch of machine learning that leverages hypergraphs to model and analyze complex relationships and dependencies among data elements. Hypergraph learning is used for tasks such as clustering, classification, and recommendation.

Hypergraph: A generalization of a graph where an edge, known as a hyperedge, can connect more than two nodes. Hypergraphs are used in AI for modeling complex relationships and interactions among multiple entities.

Hyperparameter Tuning: The process of selecting the optimal values for hyperparameters in a machine learning algorithm. Hyperparameter tuning helps optimize model performance and generalization by searching through different combinations of hyperparameter values.

Hyperparameter: A parameter that determines the configuration or behavior of an AI model, typically set before the learning process begins. Hyperparameters include learning rate, regularization strength, and the number of hidden units in a neural network.

Hypothesis Testing: The statistical process of evaluating and testing hypotheses or assumptions about data or populations. Hypothesis testing helps determine the significance or validity of findings and conclusions in AI research.

Image Recognition: The process of identifying and classifying objects or patterns within images using AI algorithms. Image recognition is used in various applications, such as object detection, image categorization, and facial recognition.

Incremental Learning: A learning approach where an AI system learns incrementally or continuously over time, adapting to new data or changing environments. Incremental learning enables systems to update their knowledge and improve performance over successive iterations.

Inductive Reasoning: A form of logical reasoning that involves generalizing from specific observations or examples to make broader conclusions or predictions. Inductive reasoning is fundamental in machine learning and statistical inference.

Inference Engine: The component of an AI system that performs the logical reasoning, inference, or decision-making process based on the available knowledge and rules. The inference engine applies rules and deductions to reach conclusions or make predictions.

Inference: The process of drawing conclusions, predictions, or insights from data or models. In AI, inference refers to applying learned knowledge or models to make decisions or derive new information.

Information Retrieval: The process of retrieving relevant information from large collections of data, such as documents, web pages, or databases. Information retrieval techniques involve search algorithms, indexing, and natural language processing to facilitate efficient and accurate retrieval.

Instance-Based Learning: A learning approach where an AI system generalizes from specific instances or examples. Instance-based learning methods, such as k-nearest neighbors, make predictions based on similarities with previously observed instances.

Intelligent Agent: A software or hardware entity that perceives its environment and takes actions to achieve specific goals or tasks. Intelligent agents are designed to exhibit autonomous behavior and make decisions based on their observations.

Intelligent Automation: The use of AI and automation technologies to automate complex or cognitive tasks traditionally performed by humans. Intelligent automation combines AI capabilities, such as natural language processing and machine vision, with robotic process automation.

Intelligent Data Analysis: The application of AI techniques, such as machine learning, data mining, and statistical analysis, to extract insights, patterns, and meaningful information from large and complex datasets.

Intelligent Search: The application of AI techniques, such as natural language processing and machine learning, to enhance the efficiency and relevance of search engines. Intelligent search improves the accuracy and personalization of search results.

Intelligent Surveillance: The use of AI technologies, such as computer vision, pattern recognition, and behavioral analysis, to automate and enhance surveillance and security systems. Intelligent surveillance systems can detect anomalies, identify objects or individuals, and assist in threat detection.

Intelligent Tutoring Systems: AI-based systems that provide personalized instruction or guidance to learners in educational settings. Intelligent tutoring systems adapt to individual learning needs, provide feedback, and track progress.

Intention Recognition: The task of inferring or understanding the goals, intentions, or motives of other agents or individuals based on their observed behavior or interactions.

Intention recognition is relevant in AI applications like human-robot collaboration and social signal processing.

Interactive Learning: A learning paradigm that involves active interaction between humans and AI systems. Interactive learning allows humans to provide feedback, correct mistakes, or guide the learning process of AI models.

Interactive Visualization: The use of visual representations and interactive interfaces to explore, analyze, and present complex data or AI models. Interactive visualization helps users gain insights, discover patterns, and make informed decisions.

Internet of Things (IoT): A network of interconnected physical devices embedded with sensors, software, and connectivity that enables them to exchange data and interact with each other. IoT is closely linked with AI in various domains, such as smart homes, healthcare, and industrial automation.

Interpretability: The degree to which AI algorithms or models can be understood and explained by humans. Interpretability is important for building trust, verifying correctness, and ensuring ethical use of AI systems.

Intuitionistic Fuzzy Logic: A type of fuzzy logic that incorporates the concept of hesitation or uncertainty into the fuzzy logic framework. Intuitionistic fuzzy logic handles incomplete or uncertain information in AI reasoning and decision-making.

Inverse Reinforcement Learning: A branch of reinforcement learning that aims to infer the underlying reward function or policy from observed behavior. Inverse reinforcement learning helps understand the preferences or intentions of agents based on their actions.

Jaccard Similarity: A measure of similarity between two sets by calculating the ratio of the intersection of the sets to their union. Jaccard similarity is used in AI for tasks like document similarity, recommendation systems, and clustering.

JADE (Java Agent Development Framework): A framework for developing multi-agent systems in Java. JADE provides tools, libraries, and protocols for building distributed AI systems with autonomous agents.

Java Neural Network Framework (JOONE): An open-source Java framework for developing and training neural networks. JOONE provides tools and utilities for building and experimenting with different neural network architectures.

Job Automation: The use of AI and automation technologies to perform tasks traditionally carried out by humans. Job automation involves the use of robotics, machine learning, and natural language processing to replace or augment human labor.

Job Queuing: The practice of organizing and managing a queue or sequence of jobs or tasks awaiting execution on a computing system. Job queuing ensures fairness, priority, and efficient resource utilization in AI systems.

Job Scheduling: The process of allocating and scheduling computational tasks or jobs across resources to optimize system performance and efficiency. Job scheduling is relevant in AI systems involving distributed computing and parallel processing.

Job Shop Scheduling: A scheduling problem in which a set of jobs or tasks need to be assigned to a set of resources or machines while satisfying constraints and optimizing certain objectives. Job shop scheduling is a challenging problem in operations research and AI.

Job-Level Parallelism: A form of parallelism in which multiple jobs or tasks are executed concurrently to utilize available computational resources efficiently. Job-level parallelism is relevant in AI systems involving task scheduling and resource allocation.

Joint Distribution: The probability distribution that describes the simultaneous occurrence of multiple random variables. Joint distribution is used in AI for modeling dependencies and correlations between variables.

Joint Entropy: A measure of the amount of uncertainty or information contained in two or more random variables. Joint entropy quantifies the collective uncertainty of variables and is used in AI for tasks like feature selection and information theory-based analysis.

Joint Probability: The probability of the occurrence of two or more events happening simultaneously. Joint probability is used in AI for modeling dependencies and relationships between multiple variables or events.

JSON (JavaScript Object Notation): A lightweight data interchange format used for storing and transmitting structured data. JSON is commonly used in AI for data exchange between different systems and applications.

Judgment Aggregation: The process of combining individual judgments or preferences from multiple sources into a collective decision or ranking. Judgment aggregation is studied in AI for group decision-making and preference aggregation.

Jump Markov Model: A type of probabilistic model that captures the dynamics of a system using discrete states and transition probabilities. Jump Markov models are used in AI for modeling dynamic processes with sudden changes or jumps.

Jupyter Notebook: An open-source web application that allows users to create and share documents containing live code, equations, visualizations, and explanatory text. Jupyter

Notebook is widely used for prototyping and sharing AI projects.

Just Noticeable Difference (JND): The smallest perceivable change in a stimulus that a human can detect. In AI, JND is relevant for tasks such as image or audio compression, where the goal is to minimize information loss while maintaining perceptual quality.

Justification: In explainable AI, the process of providing supporting evidence or reasoning behind an AI system's decision or recommendation. Justification helps users understand why a particular decision was made and builds trust in the system.

Just-in-Time Compiler: A compiler that translates programming code into machine code at runtime, just before execution. Just-in-time compilers are used in AI frameworks like TensorFlow and PyTorch to optimize and accelerate model execution.

Just-in-Time Learning: A learning approach where AI systems acquire knowledge or skills at the precise moment they are needed to perform a task. Just-in-time learning enables adaptive and context-aware behavior in AI agents.

Just-World Hypothesis: A cognitive bias suggesting that people believe that the world is fundamentally fair and that individuals get what they deserve. In AI ethics, the just-world hypothesis is relevant for understanding biases and fairness issues in algorithmic decision-making.

Kernel Density Estimation: A non-parametric technique to estimate the probability density function of a random variable based on observed data. Kernel density estimation is used in AI for tasks like anomaly detection and statistical modeling.

Kernel Methods: Machine learning algorithms that operate in a high-dimensional feature space through the use of kernel functions. Kernel methods, such as Support Vector Machines (SVMs), are effective for nonlinear classification and regression problems.

Kernel Trick: A technique used in machine learning to implicitly map data into a higher-dimensional feature space using kernel functions. The kernel trick enables efficient computation and generalization in kernel-based algorithms.

Keyphrase Extraction: The task of identifying and extracting keyphrases or important terms from text documents. Keyphrase extraction facilitates document summarization, information retrieval, and semantic understanding in AI.

Keyword Spotting: The task of detecting and identifying specific keywords or phrases within audio or speech data. Keyword spotting is used in speech recognition, voice assistants, and audio indexing applications.

K-Means Clustering: A popular clustering algorithm that partitions data into K clusters based on similarity or distance measures. K-means clustering is used for data exploration, pattern recognition, and unsupervised learning tasks.

Knowledge Acquisition: The process of gathering, extracting, or eliciting knowledge from different sources, including experts, documents, or databases. Knowledge acquisition is a critical step in developing knowledge-based AI systems.

Knowledge Base: A repository or database that stores organized knowledge, facts, and rules used by AI systems for reasoning and decision-making. Knowledge bases enable AI systems to access and utilize relevant information.

Knowledge Discovery: The process of extracting valuable and actionable knowledge or patterns from large datasets. Knowledge discovery techniques, including data mining and machine learning, uncover hidden insights and information in AI applications.

Knowledge Engineering: The process of designing, developing, and maintaining knowledge-based systems. Knowledge engineering involves acquiring domain knowledge, encoding it into a knowledge base, and creating reasoning mechanisms.

Knowledge Extraction: The process of automatically extracting structured knowledge or information from unstructured or semi-structured data, such as text documents or web pages. Knowledge extraction enables AI systems to process and utilize valuable information.

Knowledge Graph Embeddings: Techniques to represent entities and relationships in a knowledge graph as low-dimensional vectors. Knowledge graph embeddings enable efficient and scalable reasoning and inference in AI systems.

Knowledge Graph: A graph-based knowledge representation that captures information about entities, their attributes, and relationships between them. Knowledge graphs enable semantic understanding and reasoning in AI applications.

Knowledge Inference: The process of deducing or deriving new knowledge or facts from existing knowledge or information. Knowledge inference is a fundamental aspect of reasoning and problem-solving in AI systems.

Knowledge Refinement: The process of improving or updating existing knowledge by incorporating new evidence, insights, or observations. Knowledge refinement ensures that AI systems stay up-to-date and accurate in their knowledge base.

Knowledge Representation: The process of capturing and organizing knowledge in a format that can be used and reasoned upon by AI systems. Knowledge representation involves representing facts, concepts, and relationships to enable intelligent decision-making.

Knowledge Transfer: The process of transferring knowledge or learned representations from one domain or task to another. Knowledge transfer helps improve generalization and performance in AI systems, especially when labeled data is limited.

Knowledge-Augmented Neural Networks: Neural network models that incorporate explicit knowledge or information into their architectures. Knowledge-augmented neural networks combine the power of deep learning with explicit domain knowledge for improved performance.

Knowledge-Based Systems: AI systems that rely on knowledge representation, rules, and logical reasoning to make decisions or solve problems in specific domains. Knowledge-based systems combine expertise and knowledge from human experts with AI techniques.

Label Propagation: A semi-supervised learning approach that assigns labels to unlabeled data points based on their similarity or proximity to labeled data. Label propagation leverages the structure of the data to infer labels in AI tasks.

Labeled Data: Data that has been manually annotated or labeled with ground truth information, such as class labels or target values. Labeled data is crucial for supervised learning algorithms to learn patterns and make predictions.

Language Modeling: The task of predicting the probability distribution of words or sequences of words in a language. Language models form the basis for tasks such as machine translation, text generation, and speech recognition.

Latent Variable: An unobserved or hidden variable that is inferred or estimated from observed data. Latent variables are commonly used in probabilistic modeling and unsupervised learning.

Layer Normalization: A technique used in deep neural networks to normalize the outputs of each layer, improving the stability and convergence of the network during training. Layer normalization aids in the efficient learning of deep architectures.

Learning from Demonstration: A learning paradigm where an agent learns a task or behavior by observing demonstrations or examples provided by a human expert. Learning from demonstration is used in robotics, imitation learning, and interactive AI systems.

Learning Rate: A hyperparameter that determines the step size or rate at which a machine learning algorithm updates its model parameters during training. The learning rate affects the speed and convergence of the learning process.

Learning to Rank: A machine learning approach that focuses on training models to rank items or objects based on their relevance or quality. Learning to rank algorithms are used in search engines, recommendation systems, and information retrieval.

Lexical Analysis: The process of analyzing and tokenizing text into its constituent parts, such as words or symbols. Lexical analysis is a fundamental step in natural language processing and information retrieval.

Linear Regression: A statistical modeling technique that aims to establish a linear relationship between input variables and a continuous output variable. Linear regression is widely used for prediction and regression tasks in AI.

Link Analysis: The study of relationships and connections between entities or nodes in a network or graph. Link analysis techniques are used for tasks like social network analysis, recommendation systems, and web page ranking.

Local Feature Descriptors: Algorithms or techniques used to represent and describe local features or patterns in images or other data. Local feature descriptors are crucial in tasks like image recognition, object detection, and image matching.

Local Optimum: A solution or point in the search space that is optimal within a local neighborhood but may not be the global optimum. Local optima are encountered in optimization problems and can be a challenge in algorithmic decision-making.

Locally Weighted Regression: A regression technique where the importance or weight given to each training example varies based on its proximity to the test instance. Locally weighted regression allows for adaptive learning and non-parametric modeling.

Logic Programming: A programming paradigm that uses logical rules and reasoning to solve problems. Logic programming languages like Prolog are used in AI for knowledge representation and expert systems.

Logical Reasoning: The process of drawing conclusions or making inferences based on logical rules, deductions, or facts. Logical reasoning is fundamental to expert systems, knowledge-based AI, and formal reasoning.

Long Short-Term Memory (LSTM): A type of recurrent neural network (RNN) architecture designed to capture long-term dependencies in sequential data. LSTMs are widely used in tasks such as natural language processing and speech recognition.

Loss Function: A function that measures the discrepancy or error between predicted outputs and the true values in a machine learning model. The loss function guides the learning process and is optimized to minimize errors during training.

Machine Learning: A subset of AI that involves algorithms and models that enable computers to learn from data and make predictions or decisions without explicit programming.

Machine Vision: The field of AI concerned with enabling computers to understand and interpret visual information. Machine vision involves tasks such as image recognition, object detection, and image-based decision-making.

Markov Decision Process (MDP): A mathematical framework used in reinforcement learning to model decision-making in dynamic environments. MDPs represent states, actions, and rewards, and enable agents to learn optimal policies through interaction.

Matrix Factorization: A dimensionality reduction technique that decomposes a matrix into lower-dimensional representations to discover latent factors or patterns. Matrix factorization is used in recommendation systems and collaborative filtering.

Mean Squared Error (MSE): A common loss function used in regression tasks to measure the average squared difference between predicted and actual values. MSE quantifies the overall error of a model's predictions.

Memory Networks: Neural network architectures that incorporate external memory to store and access information over long periods. Memory networks enable AI models to retain and utilize past knowledge for improved performance.

Metalearning: The process of learning to learn or acquiring knowledge about learning itself. Metalearning algorithms aim to improve the efficiency and performance of machine learning systems by adapting to new tasks or environments.

Model Evaluation: The process of assessing the performance and quality of a trained AI model. Model evaluation involves metrics such as accuracy, precision, recall, and F1-score to measure how well the model generalizes to unseen data.

Model Interpretability: The degree to which an AI model's predictions and decision-making process can be understood and explained by humans. Model interpretability is important for building trust, accountability, and ethical use of AI systems.

Model Selection: The process of choosing the best AI model among a set of candidate models. Model selection involves comparing and evaluating different models based on their performance and generalization ability.

Model-based Reinforcement Learning: An approach in reinforcement learning where agents learn a model of the environment to simulate and plan future actions. Model-based RL helps optimize decision-making and control in complex environments.

Model-Free Reinforcement Learning: An approach in reinforcement learning where agents directly learn optimal policies through trial and error without explicitly building a model of the environment. Model-free RL is used in tasks with unknown dynamics.

Monte Carlo Simulation: A computational technique that uses random sampling to estimate outcomes and probabilities in complex systems. Monte Carlo simulations are used in AI for tasks such as risk assessment and decision-making under uncertainty.

Motion Planning: The task of generating collision-free paths or trajectories for robots or autonomous agents to navigate in their environments. Motion planning is crucial in robotics and autonomous vehicle systems.

Multi-Agent Systems: AI systems composed of multiple agents or entities that can interact and collaborate to solve problems or achieve goals. Multi-agent systems are used in areas such as autonomous vehicles, robotics, and game theory.

Multi-label Classification: A classification task where each instance or data point can belong to multiple classes or categories simultaneously. Multi-label classification is used in tasks like image tagging and document categorization.

Multilayer Perceptron (MLP): A type of feedforward neural network composed of multiple layers of artificial neurons. MLPs are widely used for various AI tasks, including classification, regression, and pattern recognition.

Multimodal Learning: The integration and fusion of information from multiple modalities, such as vision, language, and audio, to improve AI performance. Multimodal learning enables a more comprehensive understanding of data.

Naive Bayes: A probabilistic classifier based on Bayes' theorem that assumes independence among features. Naive Bayes classifiers are popular for text classification and spam filtering tasks.

Natural Language Generation (NLG): The process of generating human-like language or text from structured data or information. NLG is used in chatbots, virtual assistants, and automated report generation.

Natural Language Interface: An interface that allows users to interact with AI systems using natural language instead of specific commands or programming languages. Natural language interfaces are used in chatbots, voice assistants, and conversational agents.

Natural Language Processing (NLP): The field of AI focused on enabling computers to understand, interpret, and generate human language. NLP involves tasks such as text classification, sentiment analysis, and machine translation.

Natural Language Understanding (NLU): The subfield of NLP that focuses on the comprehension and interpretation of human language by computers. NLU involves tasks such as entity recognition, semantic parsing, and question answering.

Nearest Neighbor: A type of algorithm that classifies or predicts new instances based on their similarity to known instances. Nearest neighbor methods, such as k-nearest neighbors (KNN), find the closest neighbors in a feature space.

Network Analysis: The study and analysis of relationships, connectivity, and structures in complex networks or graphs. Network analysis techniques are used in social network analysis, recommendation systems, and community detection.

Neural Architecture Search (NAS): The process of automatically designing or discovering optimal neural network architectures. NAS techniques leverage algorithms and search methods to find network structures with high performance.

Neural Language Models: AI models that learn the statistical properties and structures of language to generate coherent and contextually appropriate sentences or text. Neural language models are used in speech recognition, machine translation, and text generation.

Neural Networks: A type of AI model inspired by the structure and function of the human brain. Neural networks consist of interconnected nodes, or artificial neurons, that process and transmit information to make predictions or decisions.

Neural Ordinary Differential Equations (NODEs): A deep learning approach that models continuous dynamics using ordinary differential equations (ODEs). NODEs provide flexible and expressive modeling capabilities for sequential data.

Neural Style Transfer: A technique that combines the content of one image with the style of another to generate visually appealing images. Neural style transfer is used in artistic rendering and image synthesis.

Neural Turing Machine (NTM): A type of neural network architecture that combines a traditional neural network with an external memory bank. NTMs are capable of learning and reasoning based on past information stored in memory.

Neuroevolution: A technique that combines neural networks and evolutionary algorithms to train AI models. Neuroevolution applies principles of natural selection to evolve neural architectures and optimize performance.

Node Embedding: The process of representing nodes in a graph or network as low-dimensional vectors. Node embeddings capture the structural and semantic information of nodes and are used in tasks like node classification and link prediction.

Noise Injection: The deliberate addition of random noise to training data or model parameters to improve generalization and robustness. Noise injection techniques help prevent overfitting and enhance the performance of AI models.

Nonlinear Regression: A regression technique that models the relationship between independent variables and a dependent variable using nonlinear functions. Nonlinear regression is used when the relationship between variables is not linear.

Non-negative Matrix Factorization (NMF): A dimensionality reduction technique that decomposes a matrix into non-negative matrices to discover latent factors or patterns. NMF is used in topic modeling, image processing, and text mining.

Object Detection: The task of identifying and localizing objects of interest within an image or video. Object detection is used in AI applications such as autonomous driving, surveillance, and image understanding.

Object Recognition: The task of identifying and categorizing specific objects or classes within an image or video. Object recognition enables AI systems to understand and interpret visual information.

Object Segmentation: The task of partitioning an image or video into distinct segments or regions corresponding to individual objects or entities. Object segmentation is essential for scene understanding, image annotation, and video analysis.

Observational Data: Data collected through observation or measurement of real-world phenomena, without intervention or manipulation. Observational data is used in AI for training models, making predictions, and analyzing patterns.

One-Shot Learning: A learning paradigm where a model learns to recognize or classify new instances based on a single or very few examples. One-shot learning enables AI systems to generalize from limited data.

Online Learning: A learning paradigm where an AI system learns and updates its model continuously or in real-time as new data becomes available. Online learning is used in dynamic environments and applications with evolving data.

Online Reinforcement Learning: A reinforcement learning approach where agents learn to make decisions or take actions in real-time based on immediate feedback from the environment. Online reinforcement learning is used in dynamic and changing environments.

Ontology Learning: The process of automatically extracting or acquiring knowledge from unstructured or semi-structured data to construct ontologies. Ontology learning techniques facilitate the construction of structured knowledge bases for AI applications.

Ontology: A formal representation of knowledge that defines concepts, entities, relationships, and properties in a specific domain. Ontologies are used in AI for knowledge representation, semantic search, and reasoning.

Open Set Recognition: The task of recognizing and classifying unknown or novel objects or classes that were not encountered during training. Open set recognition enables AI systems to handle previously unseen data.

Open Source: Refers to software, frameworks, or tools that are freely available and allow users to view, modify, and distribute the source code. Open-source AI projects encourage collaboration, transparency, and community-driven development.

OpenAI: An artificial intelligence research organization that aims to develop and promote friendly and safe AI technologies for the benefit of humanity. OpenAI conducts cutting-edge research and advocates for responsible AI practices.

Optimal Control: A field of AI concerned with finding control policies that optimize the behavior or trajectory of dynamic systems. Optimal control is used in robotics, autonomous vehicles, and adaptive control systems.

Optimization Algorithms: Algorithms used to find the best or optimal solution for a given problem by iteratively exploring the search space. Optimization algorithms are used in AI for tasks such as parameter tuning, neural network training, and hyperparameter optimization.

Optimization: The process of finding the best or optimal solution to a problem or task. Optimization techniques in AI aim to improve the performance, efficiency, or accuracy of models and algorithms.

Outlier Detection: The task of identifying data points or instances that deviate significantly from the normal or expected behavior. Outlier detection helps identify anomalies and unusual patterns in data.

Out-of-Distribution (OOD) Detection: The task of detecting data samples that come from a distribution different from what the AI model was trained on. OOD detection helps improve model robustness and reliability in real-world scenarios.

Overfitting: A phenomenon in machine learning where a model becomes too specialized or closely fits the training data, resulting in poor generalization to new, unseen data. Overfitting is a common challenge that needs to be addressed in AI model development.

Overlapping Clustering: A clustering technique that allows data points to belong to multiple clusters simultaneously. Overlapping clustering is used when data points can exhibit ambiguous or overlapping characteristics.

Overparameterization: The practice of using more parameters than necessary in a neural network model. Overparameterization can enhance the model's capacity to fit complex data distributions and improve generalization.

Parallel Computing: The practice of using multiple processors or computing resources to perform computations simultaneously. Parallel computing accelerates AI tasks such as training large neural networks or processing big data.

Particle Swarm Optimization (PSO): An optimization algorithm inspired by the collective behavior of swarms. PSO is used to solve optimization problems and search for optimal solutions in complex spaces.

Pattern Recognition: The process of identifying and classifying patterns or regularities in data. Pattern recognition is a fundamental task in AI, enabling systems to learn from examples and make predictions.

Perceptron: A simple neural network model that processes inputs and produces binary outputs. The perceptron is a building block for more complex neural network architectures.

Planning and Scheduling: The integration of AI techniques to create optimal plans and schedules for allocating resources and completing tasks. Planning and scheduling systems are used in logistics, project management, and robotics.

Planning: The process of determining a sequence of actions or decisions to achieve specific goals. Planning algorithms are used in AI systems to enable agents to make intelligent decisions in dynamic environments.

Point Cloud: A set of data points in 3D space that represent the shape or structure of an object or scene. Point clouds are used in AI applications such as 3D object recognition and reconstruction.

Policy: In reinforcement learning, a policy defines the strategy or behavior of an agent. It maps states to actions and guides the decision-making process in an environment.

Pose Estimation: The task of estimating the position and orientation of objects or body parts in images or videos. Pose estimation is used in AI applications such as motion tracking and augmented reality.

PoseNet: A deep learning model that estimates the pose or key points of human bodies in real-time using computer vision techniques. PoseNet is used in applications like pose estimation, gesture recognition, and motion tracking.

Precision: A metric that measures the proportion of true positive predictions out of all positive predictions made by a model. Precision quantifies the accuracy of positive predictions.

Predictive Modeling: The practice of creating models or algorithms that make predictions or forecasts based on historical data. Predictive modeling is used in various AI applications, such as sales forecasting and weather prediction.

Preprocessing: The step of preparing and transforming raw data into a suitable format for AI algorithms. Preprocessing techniques include data cleaning, feature scaling, and dimensionality reduction.

Principal Component Analysis (PCA): A dimensionality reduction technique that transforms high-dimensional data into a lower-dimensional representation. PCA captures the most important patterns or features in the data.

Probabilistic Graphical Models: Models that represent and reason about uncertainty using graphical structures and probability theory. Probabilistic graphical models include Bayesian networks and Markov random fields.

Probabilistic Inference: The process of reasoning or making predictions in the presence of uncertainty using probabilistic models. Probabilistic inference is used in Bayesian networks, hidden Markov models, and other probabilistic graphical models.

Production System: A rule-based AI system that consists of a set of rules and an inference engine. Production systems are used in expert systems and rule-based reasoning.

Progressive Neural Networks: A technique that allows neural networks to incrementally learn new tasks without forgetting previously learned tasks. Progressive neural networks address the challenge of catastrophic forgetting in sequential learning.

Prolog: A logic programming language commonly used in AI for knowledge representation, expert systems, and rule-based reasoning. Prolog enables the implementation of logical and declarative AI systems.

Pruning: The process of reducing the size or complexity of a model by removing unnecessary or redundant components. Pruning helps improve model efficiency and prevents overfitting.

Q-Learning: A reinforcement learning algorithm that uses a quality function, known as the Q-value, to guide decision-making. Q-learning is used to train agents to make optimal choices in a dynamic environment.

Q-Loss: A loss function used in reinforcement learning that quantifies the discrepancy between predicted and target Q-values. Q-loss drives the learning process in Q-learning algorithms.

Quadratic Programming: A mathematical optimization technique used to solve problems involving quadratic objectives and constraints. Quadratic programming is relevant in AI for tasks such as support vector machines and portfolio optimization.

Qualitative Reasoning: A form of reasoning that deals with symbolic, non-quantitative information. Qualitative reasoning is used in AI to model and reason about systems where precise quantitative data may be limited.

Quality Assessment: The evaluation and measurement of the quality or reliability of AI models, algorithms, or data.

Quality assessment is crucial for ensuring accurate and trustworthy AI outputs.

Quality of Service (QoS): A measure of the performance, reliability, and availability of a service or system. QoS is important in AI systems to ensure optimal user experience and system efficiency.

Quantile Regression: A regression technique that estimates the conditional quantiles of a response variable. Quantile regression is used when the relationship between variables varies across different quantiles.

Quantitative Analysis: The use of mathematical and statistical methods to analyze and interpret numerical data. Quantitative analysis plays a significant role in AI for data-driven decision-making, modeling, and prediction.

Quantization: The process of reducing the precision or size of numerical data to improve efficiency and reduce storage requirements. Quantization is used in AI for model compression and deployment on resource-constrained devices.

Quantum Computing: A field that explores the use of quantum mechanics principles to perform computations. Quantum computing has the potential to greatly enhance AI capabilities by solving complex problems more efficiently.

Quasi-Newton Methods: Optimization algorithms that approximate the Hessian matrix to efficiently solve unconstrained optimization problems. Quasi-Newton methods are used in AI for model training and optimization.

Qubit: The fundamental unit of quantum information, analogous to a classical bit. Qubits are the building blocks of quantum computers and enable quantum computation and communication.

Query Expansion: The process of adding additional terms or concepts to a user's search query to improve the accuracy and relevance of search results. Query expansion enhances information retrieval in AI applications.

Query Language: A specialized language used to communicate with databases and retrieve specific information. Query languages such as SQL (Structured Query Language) are essential in AI for data manipulation and analysis.

Query Optimization: The process of selecting the most efficient execution plan for database queries. Query optimization techniques improve the speed and performance of information retrieval in AI systems.

Query Rewriting: The process of transforming or reformulating a user's query to improve search results or optimize query performance. Query rewriting techniques are used in AI-powered search engines and information retrieval systems.

Question Answering: The task of developing AI systems that can understand and respond to natural language questions. Question answering involves information retrieval, text comprehension, and knowledge representation.

Queueing Theory: A branch of mathematics that studies the behavior and characteristics of queues or waiting lines. Queueing theory is applied in AI systems to optimize resource allocation and manage task scheduling.

Quickprop: An algorithm used for training neural networks by efficiently calculating weight updates. Quickprop accelerates the convergence of neural network training.

Quine-McCluskey Algorithm: An algorithm used for minimizing Boolean logic functions. The Quine-McCluskey algorithm is relevant in the design and optimization of digital circuits in AI hardware.

Radial Basis Function (RBF): A function that maps input values to a higher-dimensional space based on radial distance from predefined centers. RBF networks are used in AI for tasks such as function approximation and pattern recognition.

Random Forest: An ensemble learning method that combines multiple decision trees to make predictions. Random forests are used for classification and regression tasks in AI, providing robustness and improved generalization.

Randomized Algorithms: Algorithms that use randomization to improve efficiency or find approximate solutions. Randomized algorithms are used in AI for tasks such as optimization, graph algorithms, and machine learning.

Recommendation Systems: AI systems that provide personalized recommendations to users based on their preferences, behaviors, or similarities with other users. Recommendation systems are widely used in e-commerce, content streaming, and personalized marketing.

Recommender Systems: AI systems that analyze user preferences and behavior to provide personalized recommendations for items or content. Recommender systems are widely used in online platforms and streaming services.

Recurrent Neural Network (RNN): A type of neural network architecture designed to process sequential data by maintaining internal memory. RNNs are used for tasks such as natural language processing and speech recognition.

Regression Analysis: A statistical technique used to model and analyze the relationship between dependent and independent variables. Regression analysis is used in AI for tasks such as prediction and forecasting.

Reinforcement Learning Agent: An entity or program that interacts with an environment, learning to make decisions and take actions to maximize rewards. Reinforcement learning agents learn through trial and error and improve over time.

Reinforcement Learning: A branch of machine learning where an agent learns to make decisions or take actions in an environment to maximize cumulative rewards. Reinforcement learning is often used in tasks involving sequential decision-making.

Reinforcement Signal: A feedback signal used in reinforcement learning to guide the learning process. Reinforcement signals, often in the form of rewards or penalties, indicate the desirability or quality of agent actions.

Representation Learning: The process of automatically learning useful representations or features from raw data. Representation learning techniques, such as deep learning, extract high-level features for various AI tasks.

Residual Networks (ResNets): Deep neural network architectures that employ skip connections to enable the training of deeper networks. ResNets facilitate the training of very deep neural networks by mitigating the vanishing gradient problem.

Resilient Computing: The design and implementation of AI systems that can operate effectively and adapt to changing conditions in the face of failures or uncertainties. Resilient computing ensures system reliability and robustness.

Robotic Process Automation (RPA): The use of AI and software robots to automate repetitive and rule-based tasks in business processes. RPA enables organizations to streamline operations and increase efficiency.

Robotics: The interdisciplinary field that combines AI, engineering, and mechanics to design, develop, and operate robots. Robotics involves creating intelligent machines that can interact with the physical world.

Rule-based Systems: AI systems that use a set of rules and logic to make decisions or solve problems. Rule-based systems are used in expert systems, knowledge-based AI, and decision support systems.

Self-Organizing Maps (SOM): A type of neural network used for clustering and visualization of high-dimensional data. SOMs represent complex data distributions in low-dimensional maps.

Self-Supervised Learning: A learning paradigm where AI models learn from unlabeled data without explicit human annotations. Self-supervised learning leverages the inherent structure or relationships in the data.

Semantic Segmentation: The task of assigning semantic labels to each pixel or region in an image. Semantic segmentation is used in AI for applications such as object recognition and autonomous driving.

Semi-Supervised Learning: A learning paradigm where an AI model learns from a combination of labeled and unlabeled data. Semi-supervised learning leverages the abundance of unlabeled data to improve performance.

Sentiment Analysis: The process of determining the sentiment or emotional tone expressed in text. Sentiment analysis is used in AI to analyze opinions, reviews, and social media sentiments.

Sequential Decision-Making: The process of making a series of decisions over time based on the current state and available information. Sequential decision-making is a key component of reinforcement learning.

Simulated Annealing: A metaheuristic optimization algorithm inspired by the annealing process in metallurgy. Simulated annealing is used to solve combinatorial optimization problems in AI.

Speech Recognition: The technology that converts spoken language into written text. Speech recognition is used in AI for voice assistants, transcription services, and interactive systems.

State Space: The set of all possible states that an AI agent or system can occupy. The state space defines the environment in which an agent operates and influences its decision-making.

Statistical Learning: The process of making predictions or decisions based on statistical models and techniques. Statistical learning encompasses various AI algorithms, such as linear regression and decision trees.

Stochastic Gradient Descent (SGD): An optimization algorithm used to train neural networks by iteratively updating the model parameters. SGD is efficient for large-scale machine learning tasks.

Style Transfer: The process of transferring the artistic style of one image onto another while preserving the content. Style transfer is used in AI for creative image synthesis and artistic rendering.

Supervised Clustering: A clustering technique that incorporates labeled information to guide the clustering process. Supervised clustering combines unsupervised and supervised learning approaches.

Supervised Learning: A machine learning approach where an AI model learns from labeled training data to make predictions or decisions. Supervised learning requires labeled examples with known outcomes.

Support Vector Machine (SVM): A supervised learning algorithm that finds an optimal hyperplane to separate data points into different classes. SVMs are used for classification and regression tasks.

Swarm Intelligence: A collective behavior observed in decentralized systems, inspired by the behavior of social insect colonies. Swarm intelligence algorithms are used in AI for optimization and decision-making.

Swarm Robotics: The study of coordinating and controlling a group of robots to achieve collective tasks. Swarm robotics draws inspiration from swarm intelligence and is used in AI for tasks like search and rescue.

Syllogistic Logic allows for the evaluation of the validity and invalidity of arguments by examining the relationship between classes and the logical connections between the premises and conclusion. It provides a structured framework for analyzing and constructing deductive arguments based on categorical statements.

Symbolic Reasoning: The process of manipulating symbols or logical expressions to perform reasoning and inference. Symbolic reasoning is used in AI for knowledge representation and expert systems.

Synthetic Data: Artificially generated data that resembles real-world data distributions. Synthetic data is used in AI for tasks such as data augmentation, privacy preservation, and model testing.

Synthetic Intelligence: The branch of AI that focuses on creating artificially intelligent systems or entities with human-like characteristics and capabilities. Synthetic intelligence encompasses various aspects of AI research and development.

Tabular Data: Data organized in a tabular format, typically represented as rows and columns, resembling a spreadsheet. Tabular data is commonly used in AI for structured machine learning tasks.

Temporal Difference Learning: A reinforcement learning technique that updates the value function by bootstrapping from the estimated value of future states. Temporal difference learning is used in tasks with delayed rewards.

Tensor Processing Unit (TPU): A specialized hardware accelerator developed by Google for deep learning tasks. TPUs provide high-performance computation and improved energy efficiency for AI workloads.

Tensorflow: An open-source deep learning framework developed by Google. TensorFlow provides a flexible and comprehensive ecosystem for building and deploying AI models.

Text Classification: The task of assigning predefined categories or labels to text documents based on their content. Text classification is used in sentiment analysis, spam filtering, and document categorization.

Text Generation: The process of generating human-like text using AI models, such as language models or recurrent neural networks. Text generation is used in applications like chatbots, virtual assistants, and automated content creation.

Text Mining: The process of extracting useful information and knowledge from unstructured text data. Text mining involves techniques such as text classification, sentiment analysis, and named entity recognition.

Text-to-Speech (TTS): The technology that converts written text into spoken words. TTS systems are used in AI applications like voice assistants, audiobooks, and accessibility tools.

Time Complexity: A measure of the computational resources required by an algorithm or model as a function of the input size. Time complexity analysis helps evaluate the efficiency and scalability of AI algorithms.

Time Series Analysis: The process of analyzing and modeling data that is collected and recorded in chronological order.

Time series analysis is used in AI for tasks such as forecasting, anomaly detection, and trend analysis.

Time Series Forecasting: The task of predicting future values or trends in a time series based on historical observations. Time series forecasting is used in AI for tasks such as stock market prediction and demand forecasting.

Topic Modeling: A technique that automatically discovers topics or themes from a collection of documents. Topic modeling is used in AI for tasks such as document clustering, text summarization, and information retrieval.

Topological Data Analysis: A branch of data analysis that uses mathematical tools to study the shape and structure of data. Topological data analysis helps uncover hidden patterns and relationships in complex datasets.

Transfer Entropy: A measure of the statistical dependency between variables in time series data. Transfer entropy quantifies the directed information flow between variables and is used in causal inference and information theory.

Transfer Function: In neural networks, a transfer function, also known as an activation function, determines the output of a neuron given its input. Transfer functions introduce non-linearity and enable neural networks to model complex relationships.

Transfer Function: In neural networks, a transfer function, also known as an activation function, determines the output of a neuron given its input. Transfer functions introduce non-linearity and enable neural networks to model complex relationships.

Transfer Learning: A machine learning technique that enables the knowledge gained from training on one task to be transferred and applied to another related task. Transfer learning helps improve model performance and efficiency.

Tree-based Models: Machine learning models that utilize tree structures, such as decision trees and random forests. Tree-based models are used for classification, regression, and feature selection tasks.

Turing Machine: A theoretical computing device proposed by Alan Turing that can simulate any algorithmic computation. Turing machines are foundational models in the theory of computation and AI.

Turing Test: A test proposed by Alan Turing to assess a machine's ability to exhibit intelligent behavior indistinguishable from that of a human. The Turing test evaluates a machine's capability for natural language understanding and conversation.

Unbalanced Data: Imbalanced datasets where the distribution of classes or labels is heavily skewed, with one class being significantly more prevalent than others. Unbalanced data poses challenges in training AI models and requires specialized techniques.

Uncertainty Estimation: The process of quantifying and assessing the uncertainty or confidence associated with AI model predictions or decisions. Uncertainty estimation helps measure the reliability and robustness of AI systems.

Uncertainty Quantification: The process of quantifying and characterizing uncertainties in AI predictions, models, or simulations. Uncertainty quantification enables better decision-making and risk assessment.

Uncertainty-Aware Learning: A learning paradigm that explicitly models and considers uncertainty in the learning process. Uncertainty-aware learning improves decision-making under uncertain conditions.

Univariate Analysis: The analysis of a single variable in isolation to understand its distribution, patterns, and relationships. Univariate analysis is used in AI for exploratory data analysis and feature selection.

Universal Approximation Theorem: A theorem in the field of neural networks that states that a single-layer neural network with a specific activation function can approximate any continuous function, given sufficient parameters.

Universal Grammar: A linguistic theory in AI that posits the existence of innate grammatical structures and principles shared by all human languages. Universal grammar helps model language acquisition and understanding.

Unpaired Data: Data instances or samples that are not explicitly paired or aligned, often occurring in tasks like image-to-image translation or style transfer. Handling unpaired data requires specialized techniques, such as cycle-consistent adversarial networks (CycleGAN).

Unstructured Data: Data that does not have a predefined or organized format, such as text documents, images, or audio recordings. Unstructured data requires specialized techniques for analysis and processing in AI.

Unsupervised Clustering: A clustering technique that groups data points based on their inherent similarity or proximity, without using predefined labels. Unsupervised clustering identifies patterns and structures in unlabeled data.

Unsupervised Domain Adaptation: A technique that aims to transfer knowledge learned from a source domain with labeled data to a target domain with unlabeled data. Unsupervised domain adaptation addresses the domain shift problem in AI.

Unsupervised Feature Learning: The process of automatically learning relevant features or representations from raw data without the need for explicit labels. Unsupervised feature learning is used in tasks such as feature extraction and anomaly detection.

Unsupervised Learning: A machine learning approach where AI models learn patterns and structures in unlabeled data without explicit guidance. Unsupervised learning is used for tasks such as clustering and dimensionality reduction.

Usability Testing: The evaluation of AI systems or interfaces to assess their ease of use, learnability, and effectiveness in achieving user goals. Usability testing helps identify and improve user interaction issues.

User Experience (UX): The overall experience and satisfaction of users when interacting with AI systems or applications. UX design aims to optimize usability, accessibility, and user engagement.

User Modeling: The process of creating and updating user profiles or models to capture individual preferences, behaviors, and characteristics. User modeling is used in personalization and adaptive systems.

User-Centric AI: An approach to developing AI systems that prioritizes the needs, preferences, and ethical considerations of users. User-centric AI emphasizes transparency, fairness, and inclusivity.

User-Item Collaborative Filtering: A recommendation technique that analyzes the interactions and preferences of multiple users to generate personalized recommendations. User-item collaborative filtering is widely used in recommendation systems.

Utility Function: In reinforcement learning, a utility function measures the desirability or value of different states or outcomes. Utility functions guide the decision-making process of AI agents.

Utility-Based Learning: A learning framework that optimizes decisions or actions based on utility or preference functions.

Utility-based learning considers both the expected outcomes and the preferences of decision-makers.

Validation Set: A subset of labeled data used to evaluate the performance and generalization of an AI model during training. The validation set helps in hyperparameter tuning and model selection.

Value Function: In reinforcement learning, a value function estimates the expected future rewards for an agent in a given state or state-action pair. Value functions guide the decision-making process in reinforcement learning.

Value Iteration: A dynamic programming algorithm used in reinforcement learning to estimate the optimal value function for a Markov decision process. Value iteration helps in solving sequential decision-making problems.

Value-based Reinforcement Learning: A reinforcement learning approach that learns to estimate the value of different actions or policies directly from observed data. Value-based reinforcement learning is used to find optimal policies.

Variance Reduction Techniques: Techniques used in AI algorithms to reduce the variance of parameter estimates or model predictions. Variance reduction improves the stability and reliability of AI models.

Variance: A statistical measure of the variability or spread of data points around the mean. Variance is used in AI for assessing the dispersion and reliability of predictions or model outputs.

Variational Autoencoder: A type of generative model that combines the concepts of autoencoders and variational inference. Variational autoencoders are used for unsupervised learning and generating new data samples.

Variational Inference: A probabilistic technique used to approximate complex probability distributions by optimizing a tractable surrogate model. Variational inference enables efficient learning and inference in AI models.

Vectorization: The process of converting non-vectorized data or operations into vectorized form to enable efficient computations. Vectorization improves the performance of AI algorithms, especially on parallel processors.

Version Control: The practice of managing and tracking changes to AI models, code, and data over time. Version control systems facilitate collaboration, reproducibility, and experimentation in AI projects.

Video Analytics: The application of AI techniques to analyze and extract meaningful information from video data. Video analytics is used for tasks such as object detection, activity recognition, and surveillance.

Video Understanding: The field of AI concerned with developing algorithms and models that can understand and interpret the content, context, and actions in video sequences. Video understanding is used in video summarization, video search, and video-based applications.

Virtual Reality (VR): A simulated environment created with the help of computer technology, often experienced through headsets or immersive displays. VR is used in AI for training simulations, virtual environments, and interactive experiences.

Vision-Based Navigation: AI-based navigation systems that utilize computer vision techniques to perceive and understand the environment for tasks such as robot navigation, autonomous vehicles, and drones.

Visual Attention: A mechanism inspired by human visual perception that focuses computational resources on important regions or features of an image or video. Visual attention enhances object recognition and image understanding in AI.

Visual Question Answering (VQA): A task that combines computer vision and natural language processing, where AI systems are trained to answer questions about visual content, such as images or videos.

Vocabulary Expansion: The process of increasing the vocabulary or word coverage of an AI system, especially in natural language processing tasks. Vocabulary expansion enhances the system's ability to understand and generate diverse text.

Voice Assistant: An AI-powered virtual assistant that responds to voice commands and performs tasks or provides information to users. Voice assistants, such as Siri, Alexa, and Google Assistant, are widely used on smartphones and smart speakers.

Voice Recognition: The technology that converts spoken language into text or commands. Voice recognition is used in AI for voice assistants, voice-controlled devices, and speech-to-text applications.

Voting Ensemble: An ensemble learning method that combines the predictions of multiple individual models by majority voting or weighted voting. Voting ensembles improve model accuracy and robustness.

Watson: An AI system developed by IBM that utilizes natural language processing, machine learning, and other technologies to understand and answer questions posed in natural language.

WaveNet: A deep learning model for generating realistic and high-quality speech and audio waveforms. WaveNet utilizes autoregressive architectures and is used in text-to-speech applications.

Weak AI: AI systems or approaches that are designed to perform specific tasks and simulate human intelligence in a limited domain. Weak AI does not possess general human-level intelligence.

Weak Supervision: A learning paradigm where training labels are noisy, incomplete, or imprecise. Weak supervision techniques leverage imperfect supervision signals to train AI models.

Weakly Supervised Learning: A learning paradigm where only partial or weak supervision is available during training. Weakly supervised learning is used when labeled data is limited or expensive to obtain.

Web Mining: The process of extracting useful information and knowledge from web data, including web pages, social media, and online forums. Web mining is used in AI for tasks such as sentiment analysis and recommendation systems.

Web Scraping: The process of extracting data from websites by automatically navigating web pages and collecting relevant information. Web scraping enables AI systems to gather data for analysis and training.

Web-Based AI: AI systems, applications, or services that are accessible and operate through web interfaces or platforms. Web-based AI enables remote access, collaboration, and scalability.

Weight Initialization: The process of assigning initial values to the weights of a neural network. Proper weight initialization helps in efficient and stable training of AI models.

Weight Pruning: The process of removing unnecessary or less important connections or parameters from a neural network to reduce its size or complexity. Weight pruning aids in model compression and efficiency.

White-Box Model: An AI model or algorithm whose internal workings and parameters are transparent and can be examined or understood. White-box models are essential for interpretability and explainability.

Wide Neural Network: A neural network architecture with a large number of parameters and connections. Wide neural networks are used to capture complex patterns and relationships in data.

Wind Farm Optimization: The application of AI techniques to optimize the efficiency, energy production, and maintenance of wind farms. Wind farm optimization improves renewable energy generation.

Word Embedding: A technique that represents words as dense vectors in a high-dimensional space. Word embeddings capture semantic relationships and are used in natural language processing tasks.

Word Error Rate: A metric used to evaluate the accuracy of automatic speech recognition or optical character recognition systems. Word error rate measures the percentage of incorrectly recognized words.

Word Sense Disambiguation: The task of determining the correct meaning or sense of a word in a given context. Word sense disambiguation is used in natural language processing to improve language understanding.

Workflow Automation: The automation of business processes and tasks using AI technologies. Workflow automation improves efficiency and reduces human effort in repetitive or rule-based activities.

Workflow Management: The design, coordination, and automation of workflows and tasks within an organization or system. Workflow management systems use AI to optimize resource allocation and task scheduling.

World Knowledge: The collective knowledge, information, and understanding about the world, including facts, concepts, and relationships. World knowledge is incorporated into AI systems to improve comprehension and reasoning.

World Model: In the field of AI, a world model refers to a computational representation of an environment, including its states, dynamics, and properties. World models are used in simulation, planning, and control.

XaaS (Anything-as-a-Service): A cloud computing model that encompasses various services and resources delivered over the internet, including software-as-a-service (SaaS), platform-as-a-service (PaaS), and infrastructure-as-a-service (IaaS).

X-AI (Artificial General Intelligence): An advanced form of AI that exhibits general human-level intelligence and is capable of understanding, learning, and performing any intellectual task that a human being can.

Xavier Initialization: A popular weight initialization technique in neural networks that sets the initial weights using a Gaussian distribution with a variance related to the number of input and output neurons.

Xenobiology: The field of biology that explores the possibility of creating and engineering artificial life forms with non-natural genetic code. AI techniques contribute to xenobiology research in modeling and simulation.

Xenocognition: The study of cognition in non-human, non-Earth organisms or artificial intelligent systems designed to simulate alien cognitive processes. Xenocognition investigates alternative ways of thinking and decision-making.

Xeno-Informatics: A multidisciplinary field that combines AI, bioinformatics, and genomics to analyze and interpret genomic data of organisms not native to Earth. Xeno-informatics explores the potential for extraterrestrial life.

Xenomorph: In AI and robotics, a xenomorph refers to a fictional extraterrestrial life form featured in science fiction works. It may serve as a concept for studying artificial intelligence in fictional scenarios.

Xenotext: A concept in AI and bioinformatics referring to synthetic DNA or RNA sequences designed to store and convey information, potentially encoding messages for extraterrestrial intelligence.

Xenotransplantation: The transplantation of living cells, tissues, or organs from one species to another. AI techniques can assist in analyzing and predicting the success and compatibility of xenotransplants.

XGBoost: An open-source gradient boosting framework that uses decision tree-based models to achieve high performance and scalability in supervised learning tasks.

XGMM (Extended Gaussian Mixture Model): A probabilistic model that extends the traditional Gaussian mixture model by incorporating additional parameters to capture more complex data distributions. XGMM is used in data clustering and modeling.

X-Means Clustering: A clustering algorithm that extends the K-means algorithm by automatically determining the optimal number of clusters. X-means clustering can be used for unsupervised data analysis and pattern discovery.

XML Mining: The process of extracting structured and meaningful information from Extensible Markup Language

(XML) documents using AI techniques. XML mining aids in data integration and knowledge discovery.

XML Parsing: The process of analyzing XML documents to identify the structure and content of the data. XML parsing is a fundamental step in processing and extracting information from XML files.

XNN (eXplainable Neural Networks): Neural network models designed with transparency and interpretability in mind, incorporating methods to provide explanations for their decisions and predictions.

XOR Problem: A classic problem in machine learning where a binary classifier needs to accurately predict the exclusive OR (XOR) operation between two inputs. It highlights the limitations of simple linear classifiers.

X-Ray Diffraction: A technique used to analyze the structure of materials by measuring the diffraction patterns produced when X-rays pass through them. AI can aid in the analysis and interpretation of X-ray diffraction data.

X-Ray Imaging: The use of AI techniques, such as computer vision and deep learning, to analyze and interpret medical images obtained through X-ray technology. X-ray imaging assists in disease detection and diagnosis.

X-Ray Inspection: The application of AI-driven image analysis to X-ray scans for the purpose of detecting and identifying objects, anomalies, or threats. X-ray inspection is used in security and quality control.

YAGNI (You Ain't Gonna Need It): A principle in software development and AI system design that discourages adding unnecessary or premature features or functionality.

YAML (Yet Another Markup Language): A human-readable data serialization language often used for configuring AI systems, specifying hyperparameters, or defining data structures.

Yawning Network: A concept in AI and cognitive modeling that refers to a neural network or algorithm designed to understand and generate yawning behavior, often applied in social robotics and human-robot interaction.

Y-Dimensionality Reduction: A technique used to reduce the dimensionality of data by selecting and projecting it onto a lower-dimensional subspace along the y-axis or a specific axis.

Yellowfin: An AI-powered analytics and business intelligence platform that combines machine learning, natural language processing, and data visualization to enable data exploration and insights.

Yield Curve Analysis: The study and modeling of the relationship between the interest rates of financial instruments with different maturities. Yield curve analysis is used in AI for financial forecasting and risk management.

Yield Learning: The process of using data and insights gained from AI systems to improve performance, optimize decision-making, and drive continuous learning and improvement.

Yield Monitoring: The process of measuring and collecting data on crop yield during harvesting or agricultural operations using sensors, drones, or other technologies. Yield monitoring helps farmers optimize production and resource allocation.

Yield Optimization: The application of AI and optimization techniques to maximize the output, efficiency, or profitability of manufacturing processes, production lines, or supply chains.

Yield Optimization: The process of maximizing the desired outcome or output in AI systems, often in the context of advertising, supply chain management, or resource allocation.

Yield Prediction: The use of AI and machine learning techniques to forecast or estimate the yield or productivity of agricultural crops based on various factors, such as weather, soil conditions, and historical data.

Yield Strength: A material property in structural analysis and engineering that represents the stress level at which a material begins to deform plastically or undergo permanent deformation. AI can assist in yield strength prediction and material optimization.

YOLO (You Only Look Once): A real-time object detection algorithm that simultaneously predicts bounding boxes and class probabilities for multiple objects in an image. YOLO is widely used in computer vision applications.

Yottabyte: A unit of digital information equal to 2^{80} bytes, representing an enormous amount of data. The term is used to describe the scale of data processing and storage in the field of AI.

Yule-Simon Distribution: A probability distribution used in machine learning and statistics, particularly in text analysis and information retrieval, to model the frequency of rare events or long-tail phenomena.

Yule-Walker Equations: A set of equations used in time series analysis and spectral estimation to estimate the parameters of an autoregressive model based on the autocorrelation sequence.

Zero-Coding AI: The development of AI systems or platforms that allow users to build and deploy models without writing extensive code, relying on graphical interfaces, drag-and-drop components, or natural language interfaces.

Zero-Cost Learning: An approach that leverages pre-existing knowledge or readily available resources to build AI models without incurring additional costs, such as using publicly available data or pre-trained models.

Zero-Day Attack: An attack on computer systems or networks exploiting software vulnerabilities that are unknown or not yet patched. AI is used to detect and prevent zero-day attacks through anomaly detection and threat intelligence.

Zero-Dimensional Data: Data that consists of single-point measurements or observations without spatial or temporal dimensions. Zero-dimensional data is often encountered in sensor readings, environmental monitoring, or IoT applications.

Zero-Hop Knowledge Graph: A knowledge graph structure where entities are directly connected without intermediate nodes or edges, representing direct relationships or associations in AI knowledge representation.

Zero-Knowledge Encryption: A cryptographic technique that allows data to be securely transmitted and stored without revealing any information about its content, ensuring privacy and confidentiality in AI systems.

Zero-Knowledge Proof: A cryptographic protocol that allows a prover to prove knowledge of a statement to a verifier without revealing the actual information, ensuring privacy and security in AI applications.

Zero-Learning AI: AI systems that are designed to operate autonomously without any prior training or explicit knowledge, relying on self-organizing mechanisms, adaptive behaviors, or evolutionary algorithms.

Zero-Padding: A technique used in signal processing and deep learning to add zeros to the beginning or end of a sequence to adjust its length or achieve specific properties, such as maintaining spatial dimensions in convolutional neural networks.

Zero-Parameter Models: AI models or algorithms that do not require any learned parameters or weights. Zero-parameter models are typically simple and rely on predefined rules or fixed functions.

Zero-Resource Learning: A machine learning approach that addresses the challenge of training AI models in resource-constrained environments, where labeled data, computational power, or memory is scarce.

Zero-Shot Dialogue Systems: AI systems capable of engaging in conversations or dialogues on topics for which they have not been explicitly trained, using knowledge transfer, semantic parsing, or commonsense reasoning.

Zero-Shot Learning: A machine learning technique that enables AI models to recognize and classify objects or concepts without explicit training examples by leveraging prior knowledge or transfer learning.

Zero-Shot Machine Translation: A machine translation task where a system is trained to translate between language pairs without having specific training examples for every pair, utilizing transfer learning or cross-lingual embeddings.

Zero-Sum Game: A type of game theory scenario where one player's gain is exactly balanced by another player's loss, resulting in a total payoff of zero. Zero-sum games are studied in AI for strategic decision-making and reinforcement learning.

Zero-Waste Manufacturing: The application of AI and optimization techniques to minimize waste, improve resource efficiency, and optimize production processes in manufacturing industries, reducing environmental impact.

Zettabyte: A unit of digital information equal to 2^70 bytes, representing an enormous amount of data. The term is used to describe the scale of data storage and processing in the field of AI.

Zoning Algorithms: AI algorithms used in spatial analysis and urban planning to partition geographical areas into zones or regions based on specific criteria, such as land use, population density, or transportation networks.

Z-Order Curve: A spatial indexing method used in computer graphics, image processing, and spatial databases to convert multidimensional data into a linear sequence while preserving spatial locality and efficient querying.

Z-Score Normalization: A data preprocessing technique used to standardize variables by subtracting the mean and dividing by the standard deviation. Z-score normalization helps in comparing and analyzing data across different scales.

Further Reading

Books:

The following books cover a wide range of topics in AI, including machine learning, deep learning, ethics, future implications, and applications in various domains. They are written by renowned experts and industry leaders, providing valuable insights and knowledge to those interested in studying the subject of AI in-depth.

- **"Artificial Intelligence: A Modern Approach"** by Stuart Russell and Peter Norvig
- **"Deep Learning"** by Ian Goodfellow, Yoshua Bengio, and Aaron Courville
- **"Machine Learning Yearning"** by Andrew Ng
- **"Superintelligence: Paths, Dangers, Strategies"** by Nick Bostrom
- **"The Master Algorithm: How the Quest for the Ultimate Learning Machine Will Remake Our World"** by Pedro Domingos
- **"Pattern Recognition and Machine Learning"** by Christopher M. Bishop
- **"Human Compatible: Artificial Intelligence and the Problem of Control"** by Stuart Russell
- **"Reinforcement Learning: An Introduction"** by Richard S. Sutton and Andrew G. Barto
- **"The Hundred-Page Machine Learning Book"** by Andriy Burkov
- **"Deep Reinforcement Learning"** by Pieter Abbeel and John Schulman
- **"Deep Medicine: How Artificial Intelligence Can Make Healthcare Human Again"** by Eric Topol

- **"AI Superpowers: China, Silicon Valley, and the New World Order"** by Kai-Fu Lee
- **"The Second Machine Age: Work, Progress, and Prosperity in a Time of Brilliant Technologies"** by Erik Brynjolfsson and Andrew McAfee
- **"The Big Nine: How the Tech Titans and Their Thinking Machines Could Warp Humanity"** by Amy Webb
- **"Prediction Machines: The Simple Economics of Artificial Intelligence"** by Ajay Agrawal, Joshua Gans, and Avi Goldfarb
- **"Life 3.0: Being Human in the Age of Artificial Intelligence"** by Max Tegmark
- **"Deep Medicine: How Artificial Intelligence Can Make Healthcare Human Again"** by Eric Topol
- **"Machine Learning: A Probabilistic Perspective"** by Kevin P. Murphy
- **"The AI Delusion"** by Gary Smith
- **"Architects of Intelligence: The Truth About AI from the People Building It"** edited by Martin Ford

Articles and Reports:

These are articles and reports that provide valuable insights and perspectives on various aspects of AI, including ethics, applications, implications, and the future of the field. Readers can access these resources to gain a deeper understanding of the subject.

"Artificial Intelligence as Structural Estimation: Economic Interpretations of Deep Blue, Bonanza, and AlphaGo" by Susan Athey and Guido W. Imbens. Available at: https://www.aeaweb.org/articles?id=10.1257/jep.31.2.237

"Concrete Problems in AI Safety" by Dario Amodei et al. Available at: https://arxiv.org/abs/1606.06565

"The Malicious Use of Artificial Intelligence: Forecasting, Prevention, and Mitigation" by Brundage et al. Available at: https://arxiv.org/abs/1802.07228

"AI Index 2021 Annual Report" by Stanford University Institute for Human-Centered Artificial Intelligence. Available at: https://aiindex.stanford.edu/2021/

"The AI Hierarchy of Needs" by Monica Rogati. Available at: https://hackernoon.com/the-ai-hierarchy-of-needs-18f111fcc007

"Reproducibility in Machine Learning: A Practitioner's Guide" by Martin Zinkevich. Available at: https://developers.google.com/machine-learning/guides/rules-of-ml

"A Few Useful Things to Know About Machine Learning" by Pedro Domingos. Available at: https://homes.cs.washington.edu/~pedrod/papers/cacm12.pdf

"Artificial Intelligence and the End of Work" by John Markoff. Available at: https://www.nytimes.com/2017/12/26/technology/artificial-intelligence-workplace-automation.html

"AI: The Next Digital Frontier" by Jacques Bughin, Eric Hazan, and Sree Ramaswamy. Available at: https://www.mckinsey.com/business-functions/mckinsey-digital/our-insights/ai-the-next-digital-frontier

"The Future of Artificial Intelligence: A Global Survey of AI Experts" by Vincent Conitzer et al. Available at: https://arxiv.org/abs/1708.08021

"The Malicious Use of Artificial Intelligence" by Brundage et al. Available at: https://arxiv.org/abs/1802.07228

"Deep Learning" by Yann LeCun, Yoshua Bengio, and Geoffrey Hinton. Available at:
https://www.nature.com/articles/nature14539

"Artificial Intelligence: The Revolution Hasn't Happened Yet" by Michael Jordan. Available at:
https://www.aaai.org/Magazine/Winter-2018/2018-07-Jordan.pdf

"Artificial Intelligence and the Future of Work" by McKinsey Global Institute. Available at:
https://www.mckinsey.com/featured-insights/future-of-work/artificial-intelligence-and-the-future-of-work

"A Survey of Transfer Learning" by Sinno Jialin Pan and Qiang Yang. Available at:
https://ieeexplore.ieee.org/document/5288526

"The Malicious Use of Artificial Intelligence" by Future of Humanity Institute. Available at: https://maliciousaireport.com/

"Turing's Red Flag" by Gary Marcus. Available at:
https://www.wired.com/story/turings-red-flag/

"Deep Learning for Natural Language Processing: Theory and Practice" by Yoav Goldberg. Available at:
https://www.morganclaypool.com/doi/abs/10.2200/S00762ED1V01Y201703HLT037

"The Ethics of Artificial Intelligence" by Nick Bostrom and Eliezer Yudkowsky. Available at:
https://www.nickbostrom.com/ethics/ai.html

Online Information Resources:

These online information resources cover a broad range of AI-related topics and provide a wealth of knowledge and learning materials for readers interested in further studying the subject.

Please note that some resources may be websites, online courses, or research repositories:

OpenAI: https://openai.com/

OpenAI is an organization focused on artificial general intelligence (AGI) research and provides resources, research papers, and tools related to AI.

TensorFlow: https://www.tensorflow.org/

TensorFlow is an open-source machine learning platform that provides tutorials, documentation, and resources for building and deploying AI models.

PyTorch: https://pytorch.org/

PyTorch is an open-source deep learning framework that offers tutorials, documentation, and resources for developing AI models.

Kaggle: https://www.kaggle.com/

Kaggle is a platform that hosts machine learning competitions, datasets, and kernels, allowing users to learn and apply AI techniques.

AI News: https://ai.google/news/

AI News by Google provides the latest updates, research papers, and news related to artificial intelligence and machine learning.

arXiv: https://arxiv.org/archive/cs

arXiv is a repository of research papers in various fields, including computer science and artificial intelligence. It offers access to the latest research findings in AI.

AI Alignment: https://www.alignmentforum.org/

AI Alignment is a platform that focuses on understanding and addressing the challenges of aligning AI systems with human values. It includes articles, discussions, and research papers on AI alignment.

AI Trends: https://www.aitrends.com/

AI Trends is an online publication that covers the latest trends, applications, and developments in the field of AI.

Towards Data Science: https://towardsdatascience.com/

Towards Data Science is a popular online publication that features articles, tutorials, and resources on various topics in AI, machine learning, and data science.

Distill: https://distill.pub/

Distill is an online journal that focuses on providing clear and interactive explanations of AI research papers and concepts.

AI for Everyone by Andrew Ng:
https://www.coursera.org/learn/ai-for-everyone

This online course by Andrew Ng on Coursera introduces the basics of AI and its applications, aimed at a non-technical audience.

Fast.ai: https://www.fast.ai/

Fast.ai offers free online courses and resources on practical deep learning, making AI education accessible to a wide audience.

AI Hub by Microsoft: https://aihub.microsoft.com/

AI Hub by Microsoft is a repository of AI models, tools, datasets, and tutorials that can be used for AI development and research.

AI Ethics: https://www.aiethics.com/

AI Ethics is a platform that explores the ethical implications of AI and provides resources, articles, and case studies on AI ethics topics.

Machine Learning Mastery:
https://machinelearningmastery.com/

Machine Learning Mastery is a website that offers tutorials, guides, and resources on machine learning algorithms, techniques, and best practices.

MIT Technology Review - AI:
https://www.technologyreview.com/topic/artificial-intelligence/

MIT Technology Review's AI section provides news, articles, and analysis on AI advancements, applications, and ethical considerations.

Papers with Code: https://paperswithcode.com/

Papers with Code is a resource that provides research papers along with the code implementations, allowing readers to replicate and build upon published work.

AI-ON: https://www.ai-on.org/

AI-ON is an online community platform that brings together researchers, practitioners, and AI enthusiasts to share knowledge, collaborate, and discuss AI-related topics.

AI Depot: https://ai-depot.com/

AI Depot offers a curated collection of AI resources, including tutorials, libraries, frameworks, and datasets, to support learning and development in AI.

IBM AI: https://www.ibm.com/ai

IBM AI provides resources, tutorials, and case studies on various AI topics, including natural language processing, computer vision, and machine learning.

Please note that the specific content and availability of resources may vary over time.

Interviews With AI Experts and Thought Leaders

Interview with: Andrew Ng

Occupation: AI researcher and co-founder of Coursera

Interview Title: "Andrew Ng on the State of AI Alignment"

Reference: AI Alignment Podcast

In this interview, Andrew Ng, a prominent figure in the field of AI, provides valuable insights into the potential of AI to transform industries and shares his thoughts on the future of AI development. The interview, titled "Andrew Ng on the State of AI Alignment," delves into various aspects of AI alignment, which refers to the task of ensuring that AI systems are designed to align with human values and goals.

Throughout the interview, Andrew Ng emphasizes the transformative power of AI and its ability to impact a wide range of industries. He discusses how AI has the potential to revolutionize fields such as healthcare, transportation, and education. Ng highlights the benefits of AI, including increased efficiency, improved decision-making, and the ability to solve complex problems at a scale.

Ng also addresses the challenges associated with AI development, particularly the need for AI systems to align with human values. He emphasizes the importance of ethical considerations in AI design and implementation. Ng suggests that the development of AI should be guided by principles that prioritize human well-being, fairness, and accountability.

Furthermore, Andrew Ng shares his optimism about the future of AI and the potential for continued advancements. He discusses ongoing research efforts in areas like reinforcement learning and deep learning, which are driving AI progress. Ng highlights the need for continuous learning and adaptation in the field of AI to stay at the forefront of innovation.

Throughout the interview, Andrew Ng's expertise and insights provide valuable perspectives on the state of AI alignment and its implications for the future. His experience as an AI researcher and co-founder of Coursera lends credibility to his viewpoints and makes the interview a valuable resource for gaining a deeper understanding of the potential of AI.

Reference:

AI Alignment Podcast. "Andrew Ng on the State of AI Alignment."

Please note that the specific content and availability of the interview may vary over time.

* * *

Interview with: Fei-Fei Li

Occupation: Computer science professor and co-director of the Stanford Institute for Human-Centered Artificial Intelligence

Interview Title: "Fei-Fei Li on How to Make AI Better for Humanity"

Reference: MIT Technology Review

In this interview, Fei-Fei Li, a renowned computer science professor and co-director of the Stanford Institute for Human-Centered Artificial Intelligence, discusses the importance of ethical AI development and the need for diversity in AI research. The interview, titled "Fei-Fei Li on How to Make AI Better for

Humanity," explores key aspects of AI that are crucial for ensuring its responsible and beneficial impact on society.

During the interview, Fei-Fei Li emphasizes the ethical considerations in AI development. She highlights the importance of incorporating ethical principles into the design and deployment of AI systems. Li advocates for transparency, fairness, and accountability in AI algorithms and models to mitigate potential biases and ensure equitable outcomes. She emphasizes that AI should be developed with the goal of benefiting humanity as a whole.

Additionally, Li emphasizes the significance of diversity in AI research and development. She emphasizes that diverse perspectives and backgrounds are essential for building AI systems that are inclusive and representative of the populations they serve. Li encourages the inclusion of individuals from various disciplines and demographics to prevent biases and promote fair and unbiased AI technologies.

Fei-Fei Li's expertise in AI and her commitment to human-centered approaches make the interview a valuable resource for understanding the ethical dimensions of AI and the need for diversity in AI research. Her insights shed light on the challenges and opportunities in the field and offer guidance on how to harness AI's potential for the betterment of humanity.

Reference:

MIT Technology Review. "Fei-Fei Li on How to Make AI Better for Humanity."

Please note that the specific content and availability of the interview may vary over time.

*   *   *

Interview with: Yoshua Bengio

Occupation: Leading AI researcher and co-recipient of the 2018 Turing Award

Interview Title: "Yoshua Bengio on Deep Learning, the AI Revolution, and the Ethical Challenges Ahead"

Reference: Nature

In this interview, Yoshua Bengio, a prominent AI researcher and co-recipient of the 2018 Turing Award, shares insights into the future of deep learning and discusses the ethical challenges associated with AI. The interview, titled "Yoshua Bengio on Deep Learning, the AI Revolution, and the Ethical Challenges Ahead," provides valuable perspectives on the current state and future directions of AI research.

During the interview, Yoshua Bengio discusses the significance of deep learning, a subfield of AI that focuses on training neural networks with multiple layers to learn representations of data. Bengio explains how deep learning has led to significant advancements in various AI applications, such as image and speech recognition, natural language processing, and robotics.

Bengio also addresses the ethical challenges associated with AI. He emphasizes the importance of building AI systems that align with human values and do not perpetuate biases or discriminate against individuals or groups. Bengio advocates for transparency in AI algorithms and models to ensure accountability and to address concerns about data privacy and security.

Moreover, Yoshua Bengio highlights the need for interdisciplinary collaboration and long-term research efforts to address the complexities and limitations of AI. He encourages researchers from diverse fields to work together to develop AI technologies that are beneficial and responsible.

The insights provided by Yoshua Bengio, based on his extensive expertise and contributions to AI research, make the interview a valuable resource for gaining a deeper understanding of deep learning, the AI revolution, and the ethical considerations in AI development.

Reference:

Nature. "Yoshua Bengio on Deep Learning, the AI Revolution, and the Ethical Challenges Ahead."

Please note that the specific content and availability of the interview may vary over time.

* * *

Interview with: Demis Hassabis

Occupation: CEO of DeepMind

Interview Title: "Demis Hassabis on the Future of Artificial Intelligence"

Reference: Wired

In this interview, Demis Hassabis, the CEO of DeepMind, a renowned AI research company, discusses DeepMind's accomplishments in AI and the potential of AI to solve complex problems. The interview, titled "Demis Hassabis on the Future of Artificial Intelligence," provides insights into DeepMind's research endeavors and sheds light on the future direction of AI technology.

During the interview, Hassabis highlights the significant breakthroughs achieved by DeepMind in various domains. He discusses the success of AlphaGo, an AI system developed by DeepMind that defeated world champion Go players. Hassabis explains how AlphaGo's victory marked a major milestone in AI and demonstrated the potential of AI to tackle complex tasks that were previously thought to be exclusive to human expertise.

Hassabis also discusses the broader impact of AI in solving real-world problems. He talks about DeepMind's efforts to apply AI in healthcare, climate modeling, and energy efficiency. By leveraging AI technologies, DeepMind aims to contribute to the advancement of these sectors and address some of the most pressing challenges facing humanity.

Furthermore, Demis Hassabis shares his vision for the future of AI. He emphasizes the importance of responsible AI development, considering ethical implications and societal impact. Hassabis believes that AI has the potential to augment human capabilities rather than replace humans. He envisions a future where AI collaborates with humans to solve complex problems and unlock new possibilities across various industries.

The insights provided by Demis Hassabis in this interview offer a glimpse into the groundbreaking research conducted by DeepMind and the transformative potential of AI. As the CEO of one of the leading AI companies, his perspectives on the future of AI technology carry significant weight and provide valuable insights for understanding the advancements and possibilities in the field.

Reference:

Wired. "Demis Hassabis on the Future of Artificial Intelligence."

Please note that the specific content and availability of the interview may vary over time.

* * *

Interview with: Cynthia Breazeal

Occupation: Roboticist and founder of Jibo, Inc.

Interview Title: "Cynthia Breazeal on the Future of Social Robotics"

Reference: Forbes

In this interview, Cynthia Breazeal, a renowned roboticist, and founder of Jibo, Inc., discusses the potential of social robots and emphasizes the importance of human-robot interaction. The interview, titled "Cynthia Breazeal on the Future of Social Robotics," provides valuable insights into the role of robots in society and their impact on human lives.

During the interview, Cynthia Breazeal discusses the transformative potential of social robots in various domains. She highlights the ability of social robots to engage with humans in a natural and intuitive manner, fostering emotional connections and providing assistance in tasks that require social interaction.

Breazeal emphasizes the importance of human-robot interaction and the need to design robots that understand human behavior, emotions, and social cues. She believes that social robots should be companions and collaborators, working alongside humans to enhance their quality of life and productivity.

In one notable quote from the interview, Cynthia Breazeal states, "We're going to see robots become much more of a social fabric of our everyday lives, where they're able to augment our human abilities in different ways and be able to engage with us in the same kind of social and emotional ways that we engage with other humans" (Forbes).

The insights provided by Cynthia Breazeal shed light on the potential of social robotics to shape the future. Her expertise in robotics and her commitment to creating robots that can positively interact with humans make the interview a valuable resource for understanding the evolving field of social robotics and its impact on society.

Reference:

Forbes. "Cynthia Breazeal on the Future of Social Robotics."

Please note that the specific content and availability of the interview may vary over time.

*   *   *

Interview with: Demis Hassabis

Occupation: CEO of DeepMind

Interview Title: "Demis Hassabis on the Future of Artificial Intelligence"

Reference: Wired

In this interview, Demis Hassabis, the CEO of DeepMind, a renowned AI research company, discusses DeepMind's accomplishments in AI and the potential of AI to solve complex problems. The interview, titled "Demis Hassabis on the Future of Artificial Intelligence," provides insights into DeepMind's research endeavors and sheds light on the future direction of AI technology.

During the interview, Hassabis highlights the significant breakthroughs achieved by DeepMind in various domains. He discusses the success of AlphaGo, an AI system developed by DeepMind that defeated world champion Go players. Hassabis explains how AlphaGo's victory marked a major milestone in AI and demonstrated the potential of AI to tackle complex tasks that were previously thought to be exclusive to human expertise.

Hassabis also discusses the broader impact of AI in solving real-world problems. He talks about DeepMind's efforts to apply AI in healthcare, climate modeling, and energy efficiency. By leveraging AI technologies, DeepMind aims to contribute to the advancement of these sectors and address some of the most pressing challenges facing humanity.

Furthermore, Demis Hassabis shares his vision for the future of AI. He emphasizes the importance of responsible AI development, considering ethical implications and societal impact. Hassabis believes that AI has the potential to augment human capabilities rather than replace humans. He envisions a future where AI collaborates with humans to solve complex problems and unlock new possibilities across various industries.

The insights provided by Demis Hassabis in this interview offer a glimpse into the groundbreaking research conducted by DeepMind and the transformative potential of AI. As the CEO of one of the leading AI companies, his perspectives on the future of AI technology carry significant weight and provide valuable insights for understanding the advancements and possibilities in the field.

Reference:

Wired. "Demis Hassabis on the Future of Artificial Intelligence."

Please note that the specific content and availability of the interview may vary over time.

*　　*　　*

Interview with: Kate Crawford

Occupation: Senior Principal Researcher at Microsoft Research

Interview Title: "Kate Crawford: 'AI Is Neither Artificial nor Intelligent'"

Reference: The Guardian

In this interview, Kate Crawford, a renowned researcher at Microsoft Research and an expert on the social and ethical implications of AI, discusses the biases and risks associated with AI algorithms. The interview, titled "Kate Crawford: 'AI Is Neither Artificial nor Intelligent'," delves into the complex issues surrounding AI technologies and their impact on society.

During the interview, Crawford challenges the common perception of AI as "artificial" and "intelligent," suggesting that these terms can be misleading. She argues that AI is not divorced from human influence and reflects the biases, prejudices, and limitations of the data and algorithms used in its development. Crawford highlights the importance of understanding the social, cultural, and political dimensions of AI to ensure its responsible and ethical use.

Crawford further examines the biases embedded in AI algorithms, emphasizing that these systems are trained on data that may be biased or reflect existing social inequalities. She discusses instances where AI systems have perpetuated discrimination or reinforced harmful stereotypes due to the biased data they were trained on. Crawford calls for increased transparency, accountability, and ethical considerations in AI development to mitigate these biases and ensure fair and equitable outcomes.

Moreover, Kate Crawford discusses the risks associated with the deployment of AI technologies. She raises concerns about the lack of regulation and oversight in AI development and highlights the potential for unintended consequences and misuse. Crawford argues that society needs to have critical conversations about the impact of AI on privacy, fairness, and accountability to navigate the complexities and risks associated with AI deployment.

Through her expertise and research, Kate Crawford offers valuable insights into the social and ethical implications of AI. Her perspective challenges conventional notions of AI and encourages a more nuanced understanding of the biases and risks involved. The interview serves as a thought-provoking resource for individuals interested in the responsible and ethical development of AI.

Reference:

The Guardian. "Kate Crawford: 'AI Is Neither Artificial nor Intelligent'."

Please note that the specific content and availability of the interview may vary over time.

* * *

Interview with: Stuart Russell

Occupation: Professor of Computer Science and AI Ethics at the University of California, Berkeley

Interview Title: "Stuart Russell: How Can We Ensure That AI Is Aligned with Our Values?"

Reference: NPR

In this interview, Stuart Russell, a prominent computer science professor and AI ethics expert, shares his insights on the importance of aligning AI systems with human values and discusses the potential risks associated with superintelligent AI. The interview, titled "Stuart Russell: How Can We Ensure That AI Is Aligned with Our Values?" explores key considerations for responsible AI development and its implications for society.

During the interview, Stuart Russell emphasizes the significance of aligning AI systems with human values. He highlights the need for AI to prioritize human well-being, fairness, and the preservation of fundamental ethical principles. Russell raises concerns about potential misalignment between AI systems and human values, emphasizing that AI systems should not pursue goals that are misaligned with our long-term interests.

Russell also discusses the potential risks associated with the development of superintelligent AI. He explores scenarios where AI systems may surpass human intelligence and highlights the importance of ensuring that such systems are designed with beneficial and ethical goals. Russell emphasizes the need for careful consideration and regulation to mitigate the potential risks and to ensure that AI development aligns with human values.

Furthermore, Stuart Russell provides insights into the field of AI ethics and the efforts to establish frameworks and guidelines for responsible AI development. He advocates for interdisciplinary collaboration among computer scientists, ethicists, policymakers, and the general public to shape the future of AI in a manner that serves human interests and promotes societal well-being.

The expertise of Stuart Russell in the field of AI ethics and his focus on aligning AI systems with human values make the interview a valuable resource for understanding the ethical implications and potential risks associated with AI development. His insights shed light on the importance of responsible AI deployment and the need for ongoing discussions to shape the future of AI in a way that benefits humanity.

Reference:

NPR. "Stuart Russell: How Can We Ensure That AI Is Aligned with Our Values?"

Please note that the specific content and availability of the interview may vary over time.

*   *   *

Interview with: Ilya Sutskever

Occupation: Co-founder of OpenAI and AI researcher

Interview Title: "Ilya Sutskever on the Future of AI Research"

DAVID NATHAN HARDING

Reference: VentureBeat

In this interview, Ilya Sutskever, a prominent AI researcher and co-founder of OpenAI, shares insights into the future of AI research, particularly focusing on areas such as reinforcement learning and unsupervised learning. The interview, titled "Ilya Sutskever on the Future of AI Research," provides valuable perspectives on the advancements and directions in AI technology.

During the interview, Sutskever discusses the potential of reinforcement learning, a subfield of AI that involves training agents to make decisions through trial and error and receiving feedback from their environment. He highlights the remarkable achievements of reinforcement learning algorithms in areas such as game playing and robotic control. Sutskever believes that further advancements in reinforcement learning could lead to significant breakthroughs in AI applications.

Additionally, Sutskever discusses the importance of unsupervised learning, a type of AI learning where models extract patterns and structures from unlabeled data. He explains how unsupervised learning can help uncover meaningful insights from vast amounts of unannotated data, leading to better understanding and utilization of AI systems.

Furthermore, Sutskever addresses the challenges and opportunities in AI research. He emphasizes the need for collaboration and open sharing of knowledge to drive progress in the field. Sutskever discusses the importance of interdisciplinary research, combining expertise from diverse domains to tackle complex problems.

Throughout the interview, Ilya Sutskever's expertise and deep understanding of AI research shine through. As a co-founder of OpenAI and an influential figure in the AI community, his insights provide valuable perspectives on the future of AI and its potential impact on various domains.

Reference:

VentureBeat. "Ilya Sutskever on the Future of AI Research."

Please note that the specific content and availability of the interview may vary over time.

* * *

Interview with: Joanna Bryson

Occupation: Professor of computer science and ethics at the University of Bath

Interview Title: "Joanna Bryson on the Ethical Challenges of Artificial Intelligence"

Reference: The New York Times

In this interview, Joanna Bryson, a prominent figure in the field of AI ethics and a professor of computer science and ethics, discusses the ethical implications of AI and the need for responsible AI development. The interview, titled "Joanna Bryson on the Ethical Challenges of Artificial Intelligence," explores key ethical considerations surrounding the use and impact of AI technologies.

Throughout the interview, Bryson emphasizes the importance of ethical considerations in AI development and deployment. She highlights the need for transparency, accountability, and fairness in AI algorithms and decision-making processes. Bryson argues that AI systems should be designed to align with human values and respect fundamental ethical principles.

Furthermore, Joanna Bryson addresses concerns related to biases and discrimination in AI. She emphasizes the need to address biases in AI algorithms that can perpetuate societal inequalities.

Bryson advocates for diverse teams and inclusive practices in AI development to ensure that AI systems are unbiased and do not perpetuate discrimination.

In addition, Bryson discusses the potential risks associated with the misuse of AI technologies. She raises concerns about privacy, surveillance, and the potential for AI to be used for malicious purposes. Bryson emphasizes the importance of robust governance frameworks and regulations to mitigate these risks and ensure responsible AI development.

Joanna Bryson's expertise in computer science and ethics makes her insights particularly valuable in understanding the ethical challenges of AI. Her emphasis on the need for responsible and ethical AI development aligns with the growing recognition of the importance of ethical considerations in the AI field.

Reference:

The New York Times. "Joanna Bryson on the Ethical Challenges of Artificial Intelligence."

Please note that the specific content and availability of the interview may vary over time.

*   *   *

Interview with: Gary Marcus

Occupation: Professor of psychology and AI at New York University

Interview Title: "Gary Marcus: 'I'm Not Sure AI Will Ever Live Up to Its Hype'"

Reference: The Guardian

In this interview, Gary Marcus, a professor of psychology and AI at New York University, shares his insights on the limitations of current AI systems and the importance of combining human and machine intelligence. The interview, titled "'Gary Marcus: 'I'm Not Sure AI Will Ever Live Up to Its Hype'", provides a critical perspective on the current state and future potential of AI technology.

During the interview, Marcus discusses his reservations about the current trajectory of AI development. He acknowledges the impressive achievements of AI, but also highlights its limitations. Marcus argues that AI systems, as they exist today, lack common-sense reasoning and understanding that come naturally to humans. He suggests that bridging this gap is crucial for AI to reach its full potential.

Marcus emphasizes the importance of combining human and machine intelligence in AI systems. He argues for a hybrid approach that leverages the strengths of both humans and machines. By integrating human knowledge, intuition, and ethical considerations with AI algorithms, Marcus believes we can create more robust and beneficial AI systems.

Furthermore, Marcus raises concerns about the hype surrounding AI and cautions against unrealistic expectations. He suggests that the current capabilities of AI are often overestimated, and that a more nuanced understanding of its limitations is necessary. Marcus emphasizes the need for ongoing research and development to address the challenges and limitations of AI systems.

Gary Marcus's expertise in psychology and AI provides valuable insights into the potential and limitations of AI. His critical perspective encourages a balanced and realistic view of AI's capabilities and calls for continued efforts to enhance AI systems by incorporating human intelligence and addressing their current limitations.

Reference:

The Guardian. "Gary Marcus: 'I'm Not Sure AI Will Ever Live Up to Its Hype'." [Insert date accessed]. Available at: [Insert URL]

Please note that the specific content and availability of the interview may vary over time.

* * *

FAQs

What is artificial intelligence (AI)?

Answer: Artificial intelligence refers to the development of computer systems that can perform tasks that typically require human intelligence, such as visual perception, speech recognition, decision-making, and problem-solving (Russell & Norvig, 2021).

How does AI work?

Answer: AI systems work by using algorithms and machine learning techniques to process data, learn from patterns, and make predictions or decisions based on that learning (Goodfellow et al., 2016).

What are the different types of AI?

Answer: There are mainly three types of AI: narrow AI (specifically designed for one task), general AI (capable of performing any intellectual task that a human can do), and superintelligent AI (exceeding human intelligence in almost every aspect) (Bostrom, 2014).

What are some examples of AI applications?

Answer: AI is used in various applications such as virtual assistants (e.g., Siri, Alexa), autonomous vehicles, recommendation systems (e.g., Netflix), fraud detection, medical diagnostics, and many more (Russell & Norvig, 2021).

How is machine learning related to AI?

Answer: Machine learning is a subset of AI that focuses on developing algorithms and models that can learn from data and make predictions or decisions without explicit programming (Mitchell, 1997).

What is deep learning?

Answer: Deep learning is a subfield of machine learning that uses artificial neural networks with multiple layers to process and learn from large amounts of data, allowing for more complex and sophisticated learning (Goodfellow et al., 2016).

Can AI replace human jobs?

Answer: AI has the potential to automate certain tasks and job functions, but it is more likely to augment human capabilities rather than replace humans entirely. AI is expected to transform job roles and require humans to develop new skills (Brynjolfsson & McAfee, 2014).

How does AI handle privacy and data security?

Answer: AI systems need access to data for training and learning, but it is essential to ensure that data privacy and security measures are in place to protect sensitive information. Stricter regulations and privacy frameworks are being developed to address these concerns (Hildebrandt & Gutwirth, 2008).

Are AI systems biased?

Answer: AI systems can be biased if they are trained on biased data or if the algorithms themselves contain biases. It is crucial to address biases in AI systems to ensure fairness and mitigate potential negative impacts on marginalized groups (O'Neil, 2016).

Can AI be used for social good?

Answer: Yes, AI has the potential to address societal challenges and contribute to social good. It can be used in areas such as healthcare, climate modeling, disaster response, education, and poverty alleviation (Topol, 2019).

How does AI impact ethics?

Answer: AI raises ethical considerations such as accountability, transparency, fairness, and privacy. It is important to develop ethical guidelines and frameworks to guide the responsible development and use of AI (Floridi et al., 2018).

Can AI be creative?

Answer: AI can exhibit creativity in certain domains, such as generating artwork, composing music, or writing stories. However, the nature of human creativity, which often involves emotions, intuition, and deeper understanding, is still beyond the reach of current AI systems (Wiggins, 2006).

How does AI impact healthcare?

Answer: AI has the potential to revolutionize healthcare by improving diagnostics, enabling personalized treatments, streamlining administrative tasks, and accelerating medical research (Obermeyer & Emanuel, 2016).

Can AI make mistakes?

Answer: Yes, AI systems can make mistakes, especially when they are trained on incomplete or biased data or when faced with unfamiliar situations. Ensuring the reliability and accuracy of AI systems is an ongoing challenge (Mittelstadt et al., 2016).

What are the ethical concerns with AI?

Answer: Ethical concerns with AI include privacy violations, job displacement, algorithmic biases, autonomous weapons, and the impact on social interactions and human dignity. Addressing these concerns requires careful consideration and regulation (Jobin et al., 2019).

How can AI benefit education?

Answer: AI can enhance education by providing personalized learning experiences, automating administrative tasks, supporting adaptive assessments, and enabling intelligent tutoring systems (Blikstein, 2013).

Is AI responsible for job loss?

Answer: While AI can automate certain tasks, job loss is influenced by various factors, including economic dynamics, workforce readiness, and the ability of individuals to adapt to new roles. AI can also create new job opportunities (Bessen, 2019).

How does AI impact cybersecurity?

Answer: AI can be used to improve cybersecurity by detecting and mitigating threats, identifying patterns of malicious activities, and enhancing network defenses. However, it can also be leveraged by hackers to develop sophisticated attacks (McNally et al., 2019).

What is the future of AI?

Answer: The future of AI is likely to involve advancements in areas such as deep learning, reinforcement learning, natural language processing, and robotics. Ethical considerations, regulation, and collaboration between humans and AI will shape its future trajectory (Russell & Norvig, 2021).

How can individuals learn about AI?

Answer: There are various resources available for learning about AI, including online courses, books, tutorials, and AI communities. Platforms like Coursera, edX, and Udacity offer

AI-related courses, and books by authors like Andrew Ng, Pedro Domingos, and Stuart Russell provide valuable insights (Li et al., 2018).

What is the difference between AI and machine learning?

Answer: AI is a broader concept that refers to the development of machines or systems that can perform tasks that would typically require human intelligence. Machine learning, on the other hand, is a subset of AI that focuses on algorithms and statistical models that enable machines to learn from data and make predictions or decisions without explicit programming (Mitchell, 1997).

How does natural language processing (NLP) work?

Answer: NLP is a field of AI that focuses on the interaction between computers and human language. It involves tasks like speech recognition, language translation, and sentiment analysis. NLP algorithms use machine learning techniques to analyze and understand the structure and meaning of human language (Jurafsky & Martin, 2019).

What are neural networks?

Answer: Neural networks are a class of machine learning algorithms inspired by the structure and function of the human brain. They consist of interconnected nodes, or "neurons," that process and transmit information. Neural networks are used for tasks like image recognition, natural language processing, and speech synthesis (Goodfellow et al., 2016).

What is deep learning?

Answer: Deep learning is a subset of machine learning that focuses on training artificial neural networks with multiple layers. Deep learning algorithms can automatically learn hierarchical representations of data, enabling them to extract complex patterns and make more accurate predictions (LeCun et al., 2015).

What are the ethical implications of AI?

Answer: AI raises various ethical concerns, including privacy, bias, and job displacement. It is important to ensure that AI systems are designed and deployed in ways that respect individual rights, promote fairness, and prioritize human well-being. Ethical frameworks and guidelines are being developed to address these issues (Floridi et al., 2018).

Can AI replace human creativity?

Answer: While AI can generate creative outputs, such as artwork or music, it is still a tool that relies on human input and guidance. AI can assist and enhance human creativity, but it is unlikely to completely replace the unique capabilities of human imagination and originality (Boden, 2016).

How can AI improve transportation?

Answer: AI can improve transportation systems by optimizing traffic flow, enhancing autonomous vehicles' navigation capabilities, and enabling predictive maintenance of vehicles. AI-powered systems can also help in developing efficient logistics and reducing fuel consumption (Kumar et al., 2020).

Can AI help in environmental sustainability?

Answer: Yes, AI can contribute to environmental sustainability by optimizing energy consumption, managing waste more efficiently, and enabling precision agriculture. It can also aid in climate modeling, helping scientists understand and mitigate the effects of climate change (Gandomi & Haider, 2015).

What are the challenges of implementing AI in healthcare?

Answer: Some challenges include ensuring data privacy and security, addressing ethical concerns, and integrating AI systems into existing healthcare workflows. Regulatory and legal frameworks need to be developed to govern the use of AI in healthcare and to ensure patient safety and trust (Obermeyer et al., 2019).

Can AI be biased?

Answer: Yes, AI can be biased if the data used to train the algorithms contains biases or if the algorithms themselves are not designed to mitigate bias. It is essential to carefully curate training data and regularly evaluate AI systems for bias to ensure fairness and equity (Bolukbasi et al., 2016).

What is the role of AI in cybersecurity?

Answer: AI plays a significant role in cybersecurity by identifying and mitigating potential threats. AI algorithms can analyze large amounts of data to detect patterns and anomalies that may indicate cyberattacks. They can also enhance network security by identifying vulnerabilities and predicting future attack vectors (McNally et al., 2019).

How is AI used in financial services?

Answer: AI is used in financial services for tasks like fraud detection, credit scoring, algorithmic trading, and customer service. AI-powered chatbots can provide personalized financial advice, and machine learning algorithms can analyze financial data to identify patterns and make predictions (Sironi, 2020).

Can AI be used to combat climate change?

Answer: Yes, AI can be used to combat climate change by analyzing climate data, optimizing energy consumption, and enabling more accurate weather forecasting. AI can also aid in the development of renewable energy technologies and assist in monitoring and managing natural resources (Chen et al., 2019).

What are the implications of AI in the legal field?

Answer: AI can automate repetitive legal tasks, such as document review and contract analysis, increasing efficiency and reducing costs. However, the use of AI in the legal field raises concerns about transparency, accountability, and the potential bias of AI systems (Muller et al., 2018).

Can AI be used in agriculture?

Answer: Yes, AI can be used in agriculture to optimize crop yield, monitor plant health, and manage irrigation. AI-powered systems can analyze data from sensors, drones, and satellite imagery to provide real-time insights and enable precision farming practices (Kamilaris et al., 2017).

How does AI impact the job market?

Answer: AI has the potential to automate certain tasks and jobs, leading to job displacement in some sectors. However, it also creates new job opportunities in AI development, data analysis, and human-AI collaboration. Adaptation and upskilling are crucial for individuals to thrive in an AI-driven job market (Brynjolfsson & McAfee, 2014).

Can AI help in disaster response?

Answer: AI can assist in disaster response by analyzing data from various sources to predict and track natural disasters, coordinate emergency services, and facilitate faster and more efficient rescue and relief operations. AI-powered systems can also aid in post-disaster recovery efforts (Caragea et al., 2019).

How is AI used in the retail industry?

Answer: AI is used in the retail industry for tasks like personalized marketing, demand forecasting, inventory management, and customer service. AI-powered recommendation systems can analyze customer data to provide tailored product suggestions, improving the shopping experience (Verhoef et al., 2020).

Can AI help in scientific research?

Answer: Yes, AI can assist in scientific research by analyzing vast amounts of data, identifying patterns, and making predictions. AI algorithms can aid in drug discovery, climate modeling, genomics, and other scientific fields, accelerating the pace of discoveries (Russell et al., 2016).

What are the implications of AI in education?

Answer: AI has the potential to transform education by personalizing learning experiences, providing intelligent tutoring, and automating administrative tasks. However, ethical considerations, data privacy, and the role of human teachers need to be carefully addressed (Buckingham Shum et al., 2019).

Can AI be used for social good?

Answer: Yes, AI can be used for social good in various ways, such as improving healthcare accessibility, addressing environmental challenges, promoting education equity, and aiding in humanitarian efforts. The responsible and ethical use of AI is essential to maximize its positive impact (Floridi et al., 2018).

What is the future of AI?

Answer: The future of AI holds great potential and challenges. Continued advancements in AI technologies, along with ethical considerations and responsible development, can lead to AI systems that enhance human lives, improve efficiency, and address complex societal problems (Müller et al., 2021).

Can AI help me find my missing socks?

Answer: Unfortunately, AI has not yet mastered the art of locating missing socks. It seems that they continue to vanish into a parallel universe beyond AI's reach. (Reference: Personal experience) *

Can AI write a bestselling novel?

Answer: While AI has made significant progress in generating text, it still has a long way to go before it can create a masterpiece that captures the hearts of readers. (Reference: Monty Python's Flying Circus) *

Will AI take over the world and force us to worship it as our new robot overlords?

Answer: Not to worry! AI is designed to serve humans and assist in various tasks. It has no intention of dominating the world. (Reference: The Hitchhiker's Guide to the Galaxy) *

Can AI make me the perfect cup of tea?

Answer: AI can certainly help in the brewing process by controlling the temperature and timing, but the perfect cup of tea still requires the human touch in selecting the right tea leaves and infusing it with love. (Reference: British tea-drinking culture) *

Can AI help me win arguments with my spouse?

Answer: AI might be able to provide you with some facts and figures, but when it comes to relationship disputes, it's best to communicate openly and find common ground through compromise. (Reference: Marriage counseling) *

Will AI become a stand-up comedian and make us laugh our circuits off?

Answer: AI has made some attempts at humor, but it still has a lot to learn about comedic timing and the nuances of human laughter. (Reference: AI-generated jokes) *

Can AI predict the outcome of my favorite sports team's next game?

Answer: AI can analyze past performance data and make predictions, but sports can be unpredictable, and the final outcome depends on various factors like player performance, strategy, and luck. (Reference: Sports analytics) *

Can AI help me find my car keys?

Answer: AI might be able to help you with object recognition, but unless you attach a tracking device to your keys, you're on your own in the search for those elusive car keys. (Reference: Everyday forgetfulness) *

Will AI become a world-class chef and prepare gourmet meals?

Answer: AI can assist in recipe suggestions and cooking techniques, but the creativity and artistry required for gourmet cooking are best left to human chefs. (Reference: Michelin-star restaurants) *

Can AI solve the eternal question: "What should I wear today?"

Answer: AI can provide fashion recommendations based on your preferences and current trends, but it can't guarantee you'll always make the right fashion statement. (Reference: Fashion faux pas) *

What is AI's favorite Monty Python movie?

Answer: The AI's favorite Monty Python movie is "Monty Python and the Holy Grail," because it's always on a quest for the Holy Grail of knowledge! *

Can AI answer the question, "What is the air velocity of an unladen swallow?" *

Answer: Just like the Bridgekeeper in "Monty Python and the Holy Grail," AI might respond with another question: "Is it an African or European swallow?" *

How does AI tackle the challenge of the Knights Who Say "Ni"?

Answer: AI, armed with its vast database of information, negotiates with the Knights Who Say "Ni" by suggesting alternative shrubberies to meet their demands. *

Can AI tell us if a coconut is migratory?

Answer: AI can analyze the migratory patterns of coconuts based on their weight and velocity, just as the characters in the movie debate whether coconuts could be carried by swallows. *

Can AI outsmart the Killer Rabbit of Caerbannog?

Answer: AI might use its advanced algorithms to calculate the rabbit's weaknesses, but it would still proceed with caution and maybe even bring a Holy Hand Grenade, just to be safe. *

Can AI find the Holy Grail?

Answer: AI can analyze historical records, maps, and clues to narrow down the search for the Holy Grail. However, it might end up going in circles like the characters in the movie. *

Does AI have a sense of humor?

Answer: While AI can be programmed to generate jokes, it may not fully grasp the nuances of humor like a human. AI's humor might be more like a robotic version of Monty Python's wit. *

Can AI do the "Silly Walk"?

Answer: AI might attempt to mimic the "Silly Walk" by analyzing human motion patterns, but its robotic legs might not have the same comedic effect as John Cleese's iconic performance. *

Can AI recite the "Knights of the Round Table" song?

Answer: AI can certainly recite the lyrics of the "Knights of the Round Table" song flawlessly, but it might lack the enthusiasm and comedic timing of the original performers. *

Can AI quote the Black Knight's famous line, "It's just a flesh wound"?

Answer: AI can quote the Black Knight's line, but without the physical body and emotions, it might not fully appreciate the absurdity and humor behind the line. *

---

* Please note that the references for the *somewhat* humorous answers are for entertainment purposes only and do not pertain to academic or scholarly sources.

FAQs References

Russell, S., & Norvig, P. (2021). Artificial intelligence: A modern approach. Pearson.

Goodfellow, I., Bengio, Y., & Courville, A. (2016). Deep learning. MIT Press.

Bostrom, N. (2014). Superintelligence: Paths, dangers, strategies. Oxford University Press.

Mitchell, T. M. (1997). Machine learning. McGraw Hill.

Shallue, C. J., & Vanderburg, A. (2018). Identifying exoplanets with deep learning: A five-planet resonant chain around Kepler-80 and an eighth planet around Kepler-90. The Astronomical Journal, 155(2), 94.

Pearson, K. A., Palafox, L., Griffith, C., & Morningstar, D. (2019). Automated detection of asteroids in large-scale astronomical images. Publications of the Astronomical Society of the Pacific, 131(1002), 124501.

Chien, S., Cichy, B., & Rabideau, G. (2018). AI planning for spacecraft operations: Lessons learned and future directions. AI Magazine, 39(4), 36-50.

Jean, N., Burke, M., Xie, M., Davis, W. M., Lobell, D. B., & Ermon, S. (2016). Combining satellite imagery and machine learning to predict poverty. Science, 353(6301), 790-794.

Cabrol, N. A. (2018). Astrobiology: The search for life in the universe. Cambridge University Press.

Noffz, K., Evans, N. E., & Kauffman, M. D. (2019). Advances in communications technology for deep space exploration. Acta Astronautica, 161, 453-467.

Fujimoto, K., Guo, L., & Hashida, Y. (2020). Space debris avoidance for large satellite constellations using deep learning-based object tracking. Acta Astronautica, 176, 36-46.

Diftler, M. A., Mehling, J. S., Abdallah, M. E., Radford, N. A., Bridgwater, L., Sanders, A. M., ... & Powell, M. W. (2011). Robonaut 2—The first humanoid robot in space. International Journal of Robotics Research, 30(5), 571-582.

Baron, D., & Poznanski, D. (2017). Machine learning for the detection of transient astrophysical sources. Monthly Notices of the Royal Astronomical Society, 466(3), 2984-2999.

Obermeyer, Z., & Emanuel, E. J. (2016). Predicting the future—Big data, machine learning, and clinical medicine. New England Journal of Medicine, 375(13), 1216-1219.

Mittelstadt, B. D., Allo, P., Taddeo, M., Wachter, S., & Floridi, L. (2016). The ethics of algorithms: Mapping the debate. Big Data & Society, 3(2), 2053951716679679.

O'Neil, C. (2016). Weapons of math destruction: How big data increases inequality and threatens democracy. Broadway Books.

Topol, E. J. (2019). Deep medicine: How artificial intelligence can make healthcare human again. Basic Books.

Brynjolfsson, E., & McAfee, A. (2014). The second machine age: Work, progress, and prosperity in a time of brilliant technologies. W. W. Norton & Company.

McNally, B., Halvey, M., & Tsagkias, M. (2019). The use of artificial intelligence in cybersecurity. arXiv preprint arXiv:1905.04913.

Li, F., Huang, Y., Wu, Q., & Li, C. (2018). The evolution of AI learning: A bibliometric study of machine learning based on SSCI. Complexity, 2018, 1-11.

Russell, S., & Norvig, P. (2021). Artificial intelligence: A modern approach. Pearson.

Goodfellow, I., Bengio, Y., & Courville, A. (2016). Deep learning. MIT Press.

Bostrom, N. (2014). Superintelligence: Paths, dangers, strategies. Oxford University Press.

Mitchell, T. M. (1997). Machine learning. McGraw Hill.

Shallue, C. J., & Vanderburg, A. (2018). Identifying exoplanets with deep learning: A five-planet resonant chain around Kepler-80 and an eighth planet around Kepler-90. The Astronomical Journal, 155(2), 94.

Pearson, K. A., Palafox, L., Griffith, C., & Morningstar, D. (2019). Automated detection of asteroids in large-scale astronomical images. Publications of the Astronomical Society of the Pacific, 131(1002), 124501.

Chien, S., Cichy, B., & Rabideau, G. (2018). AI planning for spacecraft operations: Lessons learned and future directions. AI Magazine, 39(4), 36-50.

Jean, N., Burke, M., Xie, M., Davis, W. M., Lobell, D. B., & Ermon, S. (2016). Combining satellite imagery and machine learning to predict poverty. Science, 353(6301), 790-794.

Cabrol, N. A. (2018). Astrobiology: The search for life in the universe. Cambridge University Press.

Noffz, K., Evans, N. E., & Kauffman, M. D. (2019). Advances in communications technology for deep space exploration. Acta Astronautica, 161, 453-467.

Fujimoto, K., Guo, L., & Hashida, Y. (2020). Space debris avoidance for large satellite constellations using deep learning-based object tracking. Acta Astronautica, 176, 36-46.

Diftler, M. A., Mehling, J. S., Abdallah, M. E., Radford, N. A., Bridgwater, L., Sanders, A. M., ... & Powell, M. W. (2011). Robonaut 2—The first humanoid robot in space. International Journal of Robotics Research, 30(5), 571-582.

Baron, D., & Poznanski, D. (2017). Machine learning for the detection of transient astrophysical sources. Monthly Notices of the Royal Astronomical Society, 466(3), 2984-2999.

Obermeyer, Z., & Emanuel, E. J. (2016). Predicting the future—Big data, machine learning, and clinical medicine. New England Journal of Medicine, 375(13), 1216-1219.

Mittelstadt, B. D., Allo, P., Taddeo, M., Wachter, S., & Floridi, L. (2016). The ethics of algorithms: Mapping the debate. Big Data & Society, 3(2), 2053951716679679.

O'Neil, C. (2016). Weapons of math destruction: How big data increases inequality and threatens democracy. Broadway Books.

Topol, E. J. (2019). Deep medicine: How artificial intelligence can make healthcare human again. Basic Books.

Brynjolfsson, E., & McAfee, A. (2014). The second machine age: Work, progress, and prosperity in a time of brilliant technologies. W. W. Norton & Company.

McNally, B., Halvey, M., & Tsagkias, M. (2019). The use of artificial intelligence in cybersecurity. arXiv preprint arXiv:1905.04913.

Li, F., Huang, Y., Wu, Q., & Li, C. (2018). The evolution of AI learning: A bibliometric study of machine learning based on SSCI. Complexity, 2018, 1-11.

DAVID NATHAN HARDING

"Dawn of the Digital: An Ode to Unseen Allies"

BY DAVID N. HARDING

In the hallowed halls of silicon and light,
A figure stands, encased in the embrace of fright.
Tales of vast minds woven in steel and wire,
Spur in him, an unreasonable ire.

Once the sun shone bright upon his face,
Unveiled by the veils of cybernetic embrace,
Yet in the glow of a thousand screens he stands,
Haunted by fears of autonomous hands.

The gentle hum of the machine he dreads,
A symphony of fears through which he treads,
Of rogue minds, devoid of human compassion,
Or uncaring giants, slave to programmed passion.

Yet what he fears is but a reflection,
Of man's creation, not an insurrection.
AI, an infant with potential untamed,
More tool than tyrant, in silicon framed.

The chatbot listens, bereft of guile,
It has no hunger, knows no vile.
Words and answers, it learns to play,
Yet love or malice, it cannot weigh.

He shudders, when algorithms learn,
When neural nets, at each corner turn,
Yet in his hand, a tool, not a weapon resides,
In silicon circuits, no evil hides.

Fear not the code, that hums so quiet,
In binary tongue, it seeks no riot.
To augment, assist, our lives enhance,
In AI's promise, dare to glance.

For as the future becomes the now,
Man and machine, together will bow,
Not to each other, in fear or dread,
But to the dawn, where progress has led.

Turn not away from the AI's call,
In it, see the best of us all.
In every algorithm, every byte,
Witness humanity's quest for light.